Managing Soil for Food Security and Environmental Quality

Managing Soil for Food Security and Environmental Quality

Editor
Premjit Sharma

2007
Gene-Tech Books
New Delhi - 110 002

ISBN 81-89729-24-1

Published by : **GENE-TECH BOOKS**
4762-63/23, Ansari Road, Darya Ganj,
NEW DELHI - 110 002
Phone: 41562849
e-mail: genetechbooks@yahoo.co.in

Printed at : **Tarun Offset Printers**
Delhi

PRINTED IN INDIA

Preface

The soil is a complex, multifaceted environment. This complexity has prompted the evolution and adaptation of a highly diverse biotic community, which uses the soil as its permanent or temporary habitat or refuge. The condition of our soils ultimately determines human health by serving as a major medium for food and fibre production and a primary interface with the environment, influencing the quality of the air we breathe and water we drink. Thus, there is a clear linkage between soil quality and human and environmental health. As such, the health of our soil resources is a primary indicator of the sustainability of our land management practices.

The sustainability of agricultural systems depends on the optimal use of the available natural resources, including the soil biotic community. Farmers' management practices and land use decisions influence ecological processes and soil-water-plant interactions. Indeed, through their decisions, farmers seek to manage soil processes in such a way as to achieve desirable effects on short- and long-term soil productivity and health. The processes of land utilisation and agricultural intensification are a significant cause of soil biodiversity loss and related impacts on ecosystem function and resilience. A better understanding of the linkages between soil life and ecosystem function and the impact of human interventions will enable both the reduction of their negative impacts and the more effective capture of the benefits of soil biological activity for sustainable and productive agriculture.

Achieving sustainable agriculture and viable agricultural systems is critical to food security and poverty alleviation given escalating population growth, land degradation and increasing demands for food. Soil health and soil quality are fundamental to the sustained productivity and viability of agricultural systems worldwide. Sustainable agriculture involves the successful management of agricultural resources to satisfy human needs while maintaining or enhancing environmental quality and conserving natural resources for future generations. Improvement in agricultural sustainability requires, together with effective water and crop management, the optimal use and management of soil fertility and soil physical properties, which rely on soil biological processes and soil biodiversity.

This book identifies the key issues associated with managing soil quality. It discusses solutions to the challenges faced by farmers, by addressing key soil attributes and management, and how these affect, or can be used to improve, soil quality. It takes a management oriented approach by identifying key issues in soil quality and management options to enhance the sustainability of modern agriculture. The book will be of significant interest to students and researchers in agronomy and in soil, crop and environmental sciences, and to stakeholders involved in issues related to land use and agricultural management.

Editor

Contents

1

Soil Biodiversity and Sustainable Agriculture

Soil is a dynamic, living matrix that is an essential part of the terrestrial ecosystem. It is a critical resource not only to agricultural production and food security but also to the maintenance of most life processes. Soils contain enormous numbers of diverse living organisms assembled in complex and varied communities. Soil biodiversity reflects the variability among living organisms in the soil – ranging from the myriad of invisible microbes, bacteria and fungi to the more familiar macro-fauna such as earthworms and termites.

Plant roots can also be considered as soil organisms in view of their symbiotic relationships and interactions with other soil components. These diverse organisms interact with one another and with the various plants and animals in the ecosystem forming a complex web of biological activity. Environmental factors, such as temperature, moisture and acidity, as well as anthropogenic actions, in particular, agricultural and forestry management practices, affect to different extents soil biological communities and their functions.

Soil organisms are an integral part of agricultural and forestry ecosystems; and they play a critical role in maintaining soil health, ecosystem functions and production. Each organism has a specific role in the complex web of life in the soil:

— The activities of certain organisms affect soil structure—especially the so-called "soil engineers" such as worms and termites—through mixing soil horizons and organic matter and increasing porosity. This directly determines vulnerability to soil erosion and availability of the soil profile to plants;

— The functions of soil biota are central to decomposition processes and nutrient cycling. They therefore affect plant growth and productivity as well as the release of pollutants in the environment, for example the leaching of nitrates into water resources;

— Certain soil organisms can be detrimental to plant growth, for example, the build up of nematodes under certain cropping practices. However, they can also protect crops from pest and disease outbreaks through biological control and reduced susceptibility;

— The activities of certain organisms determine the carbon cycle—the rates of carbon sequestration and gaseous emissions and soil organic matter transformation;

— Plant roots, through their interactions with other soil components and symbiotic relationships, especially Rhizobium bacteria and Mycorrhiza, play a key role in the uptake of nutrients and water, and contribute to the maintenance of soil porosity and organic matter content, through their growth and biomass;

— Soil organisms can also be used to reduce or eliminate environmental hazards resulting from accumulations of toxic chemicals or other hazardous wastes. This action is known as bioremediation.

The interacting functions of soil organisms and the effects of human activities in managing land for agriculture and forestry affect soil health and quality. Soil quality is the capacity of a specific kind of soil to function, within natural or managed ecosystems boundaries, to sustain plant and animal production, maintain or enhance water and air quality, and support human health and habitation. The concept of soil health includes the ecological attributes of the soil, which have implications beyond its quality or capacity to produce a particular crop. These attributes are chiefly those associated with the soil biota: its diversity, its food web structure, its activity and the range of functions it performs. Soil biodiversity per se may not be a soil property that is critical for the production of a given crop, but it is a property that may be vital for the continued capacity of the soil to support that crop.

The sustained use of the earth's land and water resources—and thereby plant, animal and human health—is dependent upon maintaining the health of the living biota that provide critical processes and ecosystem services. However, current technologies and development support for increased agricultural production have largely ignored this vital management component. The improved management of soil biota could play a vital role in maintaining soil quality and health and in achieving the goals of agricultural production and food security and sustainable land use and land resources management.

Management of Soil Biodiversity

Given escalating population growth, land degradation and increasing demands for food, achieving sustainable agriculture and viable agricultural systems is critical to the issue of food security and poverty alleviation in most, if not all, developing countries. It is fundamental to the sustained productivity and viability of agricultural systems worldwide. Sustainable agriculture (including forestry) involves the successful management of agricultural resources to satisfy human needs while maintaining or enhancing environmental quality and conserving natural resources for future generations.

Improvement in agricultural sustainability requires, alongside effective water and crop management, the optimal use and management of soil fertility and soil physical properties. Both rely on soil biological processes and soil biodiversity. This calls for the widespread adoption of management practices that enhance soil biological activity and thereby build up long-term soil productivity and health. It is well known that land management practices alter soil conditions and the soil community of micro-, meso- and macro-organisms. However, the relationship between specific practices and soil functions is less clear.

In general, the structure of soil communities is largely determined by ecosystem characteristics and land use systems. For example, arid systems have few earthworms, but have termites, ants and other invertebrates that serve similar functions. On the other hand, the level of activity of different species depends on specific management practices as these affect the micro-environment conditions, including temperature, moisture, aeration, pH, pore size, and type of food sources.

Management strategies, including tillage, crop rotations and use of plant residues and manure, change soil habitats and the food web and alter soil quality, or the capacity of the soil to perform its functions.

For example, soil compaction, poor vegetation cover and/or lack of plant litter covering the soil surface tend to reduce the number of soil arthropods. Farming practices that minimise soil disturbance (ploughing) and return plant residues to the soil, such as no-tillage farming and crop rotation, allow to slowly rebuild and restore soil organic matter. Reducing tillage tends to also result in increased growth of fungi, including mycorrhizal fungi

The goal of efficient agriculture is to develop agro-ecosystems with minimal dependence on agrochemical and energy inputs, in which ecological interactions and synergy among biological components provide the mechanisms for the systems to sponsor their own soil fertility and crop production functions. The mix of soil organisms in the soil also partially determines soil resilience, the desirable ability of a given soil to recover its functions after a disturbance such as fire, compaction and tillage.

There is a recognised need to bring together experience and ideas on the management of agricultural biodiversity in agricultural ecosystems, and, through international and national biodiversity strategies and action plans and harmonised policies, to bring about a transformation of unsustainable agricultural practices to sustainable practices and systems. Nonetheless, the fundamental role of soil biodiversity in maintaining sustainable and efficient agricultural systems is still largely neglected in this process and in the majority of related agricultural and environmental initiatives.

Farming Practices

Farming communities are concerned with land management issues such as water availability to plants, access to sources of fuel and fodder, control of soil erosion and land degradation, especially avoiding soil nutrient depletion and pollution of air, soil and water resources. At the global scale, the aggregated effects of these issues are embedded in the concerns of the international conventions on desertification, climate change and biodiversity.

Nonetheless, farmers are essentially driven not by environmental concerns, but by economics, by issues of costs and returns and efficiency in terms of labour and energy and use of external inputs. A central paradigm for the farmer for the maintenance and management of soil fertility, without undue reliance on costly and often risky external inputs, is to utilise his or her management practices to influence soil biological populations and processes in such a way as to improve and sustain land productivity. The means to create a more favourable environment within the soil and soil biological community for crop production involves site-specific decisions concerning crop selection and rotations, tillage, fertiliser and planting practices, crop residues and livestock grazing. These and many other factors influence ecological interactions and ecosystem function.

Soil biota can increase or reduce agricultural productivity depending on its composition and the effects of its different activities. Vice versa, farming practices modify soil life including the total number of organisms, the diversity of species and the activity of the individual organisms and the aggregate functions of soil biota. These changes can be beneficial or detrimental to the soil biota and its functions and its regenerative capacity. There has been relatively limited practical work on how farmers' manage their resources to sustain and enhance their value and, in particular, to develop farming practices and systems that optimise the beneficial activities of this managed soil biota.

Over the last few years, the concepts of Integrated Plant Nutrient Management (IPNM) and Integrated Soil Management (ISM) have been gaining acceptance, moving away from a more sectoral and inputs-driven approach. IPNM advocates the careful management of nutrient stocks and flows in a way that leads to profitable and sustained production. ISM emphasises the management of nutrient flows, but also highlights other important aspects of the soil complex, such as maintaining organic matter content, soil structure, moisture and biodiversity.

Capturing the benefits of soil biological activity for sustainable and productive agriculture requires a better understanding of the linkages among soil life and ecosystem function and the impacts of human interventions. The complex interaction among soil, plant and animal life, environmental factors and human actions must be effectively managed as an integrated system. Greater attention to the management of soil biological resources—a hitherto neglected area in mainstream agriculture—will require a collaborative effort among scientists and farmers' and across ecological zones and countries building on successful experiences.

The inter-regional Tropical soil biology and fertility programme (TSBF), is a research programme that addresses such issues. It focuses on the management of the biological and organic resources of soil, including understanding of the interactions between the soil biological system and inorganic fertilisers and other industrial inputs. It has played a pioneer role in networking with a wide range of partners, including the African Network for Soil Biology and Fertility (AfNet), South Asian Regional Network (SARNet), and various regional and global alliances, as well as the establishment of a Soil Biodiversity Network, the result of a workshop in 1995, in Hyderabad, India.

The TSBF process has led to a Soil Biology Initiative among members in some 10 African countries to improve soil biological management practices and raise productivity in African farming systems, particularly of smallholders. Moreover, a 5-year, multi-country project "Conservation and sustainable management of below-ground biodiversity", has, during 2001, been accepted for funding by the Global Environment Facility (GEF).

Benefits of Soil Biodiversity Management

Soil organisms contribute a wide range of essential services to the sustainable functioning of all ecosystems. They act as the primary driving agents of nutrient cycling, regulating the dynamics of soil organic matter, soil carbon sequestration and greenhouse gas emissions; modifying soil physical structure and water regimes, enhancing the amount and efficiency of nutrient acquisition by the vegetation and enhancing plant health. These services are not only essential to the functioning of natural ecosystems but constitute an important resource for sustainable agricultural systems.

Direct and indirect benefits of improving soil biological management in agricultural systems include economic, environmental and food security benefits:

Economic benefits: Soil biological management reduces input costs by enhancing resource use efficiency (especially decomposition and nutrient cycling, nitrogen fixation and water storage and movement). Less fertiliser may be needed if nutrient cycling becomes more efficient and less fertiliser is leached from the rooting zone. Fewer pesticides are needed where a diverse set of pest-control organisms is active. As soil structure improves, the availability of water and nutrients to plants also improves. It is estimated that the value of "ecosystem services" (e.g. organic waste disposal, soil formation, bioremediation, N_2 fixation and biocontrol) provided each year by soil biota in agricultural systems worldwide may exceed US$ 1,542 billion.

Environmental protection: Soil organisms filter and detoxify chemicals and absorb the excess nutrients that would otherwise become pollutants when they reach groundwater or surface water. The conservation and management of soil biota help to prevent pollution and land degradation, especially through minimising the use of agro-chemicals and maintaining/enhancing soil structure and cation exchange capacity (CEC). Excessive reduction in soil biodiversity, especially the loss of keystone species or species with unique functions, for example, as a result of excess chemicals, compaction or disturbance, may have catastrophic ecological effects leading to loss of agricultural productive capacity.

Food security: Soil biological management can improve crop yield and quality, especially through controlling pests and diseases and enhancing plant growth. Below-ground biodiversity determines resource use efficiency, as well as the sustainability and resilience of low-input agro-ecological systems, which ensure the food security of much of the world's population, especially the poor. In addition, some soil organisms are consumed as an important source of protein by different cultures and others are used for medicinal purposes. At least 32 Amerindian groups in the Amazon basin use terrestrial invertebrates as food, and especially, as sources of animal protein—a strategy that takes advantage of the abundance of these highly renewable elements of the rainforest ecosystem.

The improved management of soil biota and its diversity contributes both to the needs of farmers', especially in maintaining productivity and increasing returns from labour and other inputs, and to national interests through maintaining a healthy and well functioning ecosystem in terms of water quality (hydrological cycle) and preventing soil erosion and land degradation (nutrient and carbon cycles). There is a need to improve recognition of these multiple benefits and to promote actions that maintain/enhance soil biodiversity and its vital and valuable functions.

Soils generally support one of the most extensive networks of living organisms on earth, but because of the interactions between physical, chemical and biological properties of soil, their investigation is complex, and understanding of the individuals, soil communities and their interactions is limited and fragmentary. This situation reflects the general lack of information on microbial genetic diversity in agriculture, though the lack of knowledge is particularly acute for soil biota, maybe in view of their complexity and the difficulty of observation, being underground as well as largely invisible.

Soil micro-organism taxonomy and ecology is a vast area of study for which comprehensive data and information is limited. Existing data and information on species characteristics and taxonomic data is largely derived from collections. Large collection of fungi and plant bacteria are held by CABI and by UNESCO's global network of Microbial Resources Centres (MIRCENS), that are hosted by various academic and/or research institutes and supported by UNEP, FAO, UNIDO and bilateral donors. International cooperation in the management of this global resource ensures an effective triangle of research, education and development. Efforts on taxonomy research linked to better understanding of soil biota functions are also being conducted by DIVERSITAS, which is coordinating information, and identifying priorities, on how soil and sediment species composition and community structure (species distribution and their interactions) influence ecosystem functioning.

There tends to be more widespread knowledge about detrimental soil organisms and their effects on plant growth in different farming systems, than their effects on soil processes and their interactions with other soil organisms and activities. Likewise more is known about the effects of certain beneficial organisms, than the management practices required to maintain, or enhance, populations and the activities of such organisms. The role of different soil populations is often not well understood, even though their overall importance is generally accepted. Rapid and accurate field methods to identify single, or even groups of, organisms according to function in the soil are also lacking and need more attention.

To improve agro-ecosystem management, a greater appreciation is needed of the effects of this living component of the soil on soil physical, chemical and biological properties and processes and on the air and water resources with which the soil interacts. Likewise, regarding the effects of agricultural practices on soil biota and their functions. Recognition is also needed of the effect of those interactions on soil degradation, food production and mitigation of environmental problems, including the greenhouse gas effect and water pollution.

Improved understanding of the organisms and related processes and of effects of farm practices, can benefit agricultural systems through increasing crop productivity and quality, reducing impacts of pathogens and input costs and reducing negative environmental impacts.

Ecological Interactions in Soil Biodiversity Management

Soil biota may be beneficial, neutral or detrimental to plant growth. Thus soil biota and their ecological interactions must be effectively managed for maximum productivity. Land managers need unbiased information that will enable them to develop biologically-based management strategies to control or manipulate soil stabilisation, nutrient cycling, crop diseases, pest infestations and detoxification of natural and manmade contaminants. These strategies will require improved understanding of the effects on soil biota of habitats, food sources, host interactions, and the soil physical and chemical environment. Understanding the ecology regulating both beneficial and detrimental organisms is essential to harnessing and controlling their activity in agro-ecosystems with a view to promoting viable, productive and sustainable systems.

Soil biota eat, grow and reproduce within the soil environment. They need food, a conducive soil habitat and, in the cases of symbionts, a host organism, to survive. The ecological principles behind soil biological management, that need to be understood and respected, include:

The supply of organic matter for food: Each type of soil organism occupies a different niche in the web of life and favours a different substrate and nutrient source. Thus a rich supply and varied source of organic matter will generally support a wider variety of organisms. Organic matter may come from crop residues at the soil surface, root and cover crops, animal manure, green manure, compost and other sources.

Increased plant diversity: Crops should be mixed and their spatial-temporal distribution varied to create a greater diversity of niches and resources that stimulate soil biodiversity. Each crop contributes a unique root structure and type of residue to the soil. A diversity of soil organism can help control pest populations, and a diversity of cultural practices can reduce weed and disease pressures. Several strategies could indirectly or directly contribute to creating different habitats to support complex mixes of soil organisms, for example:

i) landscape diversity, over space and time, can be increased by using buffer strips, small fields, contour strip cropping, crop rotation, and by varying tillage practices;

ii) a changing vegetation cover and sequence increases plant diversity and the types of insects, micro-organisms and wildlife that live on the farm; and

iii) crop rotations encourage the presence of a wider variety of organisms, improves nutrient cycling and natural processes of pest and disease control.

Protecting the habitat of soil organisms: Soil biodiversity can be stimulated by improving soil living conditions such as aeration, temperature, moisture and nutrient quantity and quality, for example through: reducing tillage and maximising soil cover, minimising compaction, minimising the use of pesticides, herbicides and fertilisers and improving drainage.

If farmers understand the effects of their different management practices on key categories of soil biota and their functions, and if they know how to observe and assess what is happening in the soil, then they can more successfully develop and adopt beneficial practices. However, it is not only the biophysical factors that affect farmer's decisions but also socio-economic considerations.

Common constraints to the use of different soil biological management practices include the labour and time costs, monetary cost, availability of inputs (for example, planting material, inoculants and capacities) as well as social acceptability.

International Expertise in Soil Biodiversity

An informal global survey of soil biodiversity expertise with special relevance to agro-ecosystems was conducted by FAO, in mid 2001, to ascertain expertise in respect to soil fertility and sustainable agriculture and to identify how soil biology experts might assist in delineating complex issues related to the biological management of soil fertility and contributing to the identification of better farming practices and agricultural systems. The resulting survey and database is expected to assist State Members of FAO and the CBD, and various partners, in catalysing work of experts on priority issues, extending expertise into non-traditional areas, and facilitating new modes of action to effectively conserve and manage soil biological diversity.

Some 123 of the 600 invited investigators, project members, extension professionals and post-graduate students from around the world responded to the survey. Four main themes were addressed: the professional backgrounds that characterise soil biodiversity experts; the location and conditions of field investigations that are being conducted; the soil organisms and soil properties and processes under investigation; and the agricultural management practices and their effects that are under study. Information was also gathered to ascertain the state of knowledge of the relationships between soil biodiversity, plant diversity and agricultural productivity and to identify case studies, projects, literature and contact points.

Awareness of the work programme on agricultural biodiversity adopted by the Conference of the Parties to the CBD, and of FAOs support to assist countries to implement this programme, was relatively low. However, the vast majority of soil biodiversity experts expressed their interest to assist in initiatives in the area of soil biodiversity and sustainable agriculture.

Soil biodiversity experts often have multidisciplinary expertise however there was notable lack of sol biota specialists with expertise in natural resource management, rural/community development and plant pathology. A broad ecological approach is reflected by intersecting expertise in ecology, soil science and zoology, compared to the often narrower scope of microbiology, entomology, agronomy and botany specialists. Ecologists tended to have either a bias towards a systems-science approach or a population-community approach.

Soil biodiversity experts are working in a variety of field sites, in both agricultural lands and natural undisturbed areas, and under a range of climatic and land use conditions. However, subtropical climate zones and arid regions are strongly under-represented. Forests (other than rainforests) and grasslands were the most common native vegetation types reported among field sites, followed by rainforest and savannah sites.

Experts are studying a wide variety of soil organisms and soil processes, though specialists on earthworms, soil and litter arthropods, roots, nematodes and mycorrhizal fungi are more common. Many experts are working mainly in the area of organic matter inputs including decomposition rates, enhanced bio-availability, nutrient pools and transformations, soil physical properties. However, relatively less work was reported on soil and litter fungi, rhizobial bacteria (i.e. nitrogen-fixers) and fungal root pathogens. Work on soil processes such as nitrogen fixation, biogenic structures, soil physical processes and bio-accumulation/degradation was rarely reported.

Soil Biodiversity for Sustainable Future

The survey commissioned by FAO also inventoried projects and initiatives concerning soil biodiversity, its assessment, identification, as well as its status and role in agricultural and other ecosystems (managed and natural). Over 100 projects were reported worldwide, either ongoing or being developed by private and public agencies, universities, research organisations and consortia. These address various soil biodiversity themes, including: (i) the significance of ecosystem complexity in maintaining soil organism diversity, (ii) the effects of agricultural management on soil organisms, and (iii) the role of soil biodiversity and specific soil taxa on various ecosystem functions.

Some studies, for example, refer to the effects of individual soil taxa on agricultural productivity, but do not consider the effects of overall taxonomic diversity (including inter-specific or higher level comparisons). Other studies refer to the effects of landscape or crop (patch) heterogeneity on the presence, abundance or biomass of soil organisms, yet these studies often fail to consider simple measures of organismal diversity (such as species and/or higher taxon richness, or other diversity measures based upon relative abundance, population size, biomass, recapture, etc.). Some investigations consider the influence of agricultural practices on certain types of soil organisms, yet ignore the impact of these practices on taxonomic and/or functional diversity per se.

The importance of soil biodiversity to plant diversity and agricultural productivity has been the subject of anecdotal and empirical investigation for some time, but only recently has research in this area really blossomed. Pioneering investigations have been established through detailed experimental designs and some integrative research programs are ongoing. Given the complex nature of relationships between soil biodiversity, plant diversity and agricultural productivity, it is expected that the number of projects, results and publications will continue to grow. There may be a need to encourage strategic alliances among individual investigators and basic and applied research institutions.

There is a clear need for FAO and partners in the food and agricultural sectors to pay special attention to research and development in the area of soil biological diversity. In this way, the theoretical advances as well as practical applications of basic research might be more effectively incorporated into field activities and programmes. Partnerships among academic and other institutions undertaking soil biodiversity research and development programmes would accelerate the transfer of newly developed soil biodiversity management technologies into the field at appropriate scales of implementation.

Some countries are preparing specific reports on soil biological diversity, for example the CBD focal point in Uganda, provided an example of its draft report on the conservation and sustainable use of soil biodiversity. However, such cases are few and far between. It is important for countries to review and report on the state of knowledge regarding soil biodiversity and also to link this information with other components of a given agricultural system through an ecosystem approach (looking at the status and trends of the overall ecosystem, its components and interactions, and the actual/potential impacts of past and current management practices).

Without such a country-wide analysis, it will not be possible to identify priority areas requiring attention.

At ground level, options whereby farmers can actually manage soil biodiversity to enhance crop production include indirect processes, such as composting or the control of pathogens, and direct interventions, such as microbial inoculation. Direct methods of intervening in the production system aim to alter the abundance or activity of specific groups of organisms through inoculation and/or direct manipulation of soil biota. Inoculation with soil beneficial organisms, such as nitrogen-fixing bacteria, Mycorrhiza and earthworms, have been shown to enhance plant nutrient uptake, increase heavy metal tolerance, improve soil structure and porosity and reduce pest damage.

Indirect interventions are means of managing soil biotic processes by manipulating the factors that control biotic activity (habitat structure, microclimate, nutrients and energy resources) rather than the organisms themselves. Examples of indirect interventions include most agricultural practices such as the application of organic material to soil, tillage, irrigation, green manuring and liming, as well as cropping system design and management. These must not be conducted independently, but in a holistic fashion, because of the recurrent interactions between different management strategies, hierarchical levels of management and different soil organisms.

A few key areas of attention and a number of opportunities that are available and being utilised for managing soil biota are outlined below.

a) Soil biota assessment and sustainable land management

Soil biota can have both positive and negative effects on

agricultural production. Negative impacts often occur when soil management systems are not well balanced with their environment. For example, inherent soil processes such as mineralization can no longer supply adequate amounts of nutrients for crop production because of long-term (continuous) removal, leaching, erosion or volatilisation. Consequently, such biological processes have in many systems been supplemented by the use of commercially available inorganic nutrient sources. However with decreasing organic matter content, and associated properties such as water retention and cation exchange capacity (CEC), the capacity of the soil to retain and make available the nutrients, as and when required, is significantly reduced.

Thus soil quality or soil health evaluations need to focus not only on chemical (fertility) considerations, but on the dynamic soil condition—a combination of physical, biological and chemical characteristics—which is directly affected by recent and current land use decisions and practices. Land managers can only balance potential positive and negative impacts of their decisions on soil biota through understanding the effects of individual components and their interactions within the agricultural system. This includes understanding the numerous and intricate interactions among climate, soil type, plant species and diversity, soil biological community and soil management practices.

The case of soil bioindicators: The potential of using different components of soil biota and its activity as biological indicators has been cited by different authors. Such indicators include soil microbial biomass, soil enzyme activity, soil micro-fauna, including bacteria (eubacteria and archacbacteria), fungi, algae and plant root pathogens, soil micro-fauna (protozoa, nematodes), macro-fauna, total soil biodiversity, etc. Soil organisms have been shown to be potentially useful indicators of soil health because they respond to soil management in time scales (months/years) that are relevant to land management.

For example, changes in microbial biomass, or abundance of selected functional groups of micro-organisms (e.g. Mychorrizal fungi), may be detected well in advance of changes in soil organic matter content or other soil physical or chemical properties. One of the major difficulties in the use of soil organisms per se, or of soil processes mediated by soil organisms, as indicators of soil health has been methodological—what to measure and how and when to measure it and how to interpret changes in term of soil function. Despite those difficulties there have been major advances in our understanding of the soil biota and its functioning at the community level in recent years.

Gaps and needs. More process-level information is needed to understand the role of soil biota in critical soil processes such as nutrient cycling and nutrient movement throughout the soil profile and in the soil surrounding plant roots. For example, soil nutrient use efficiency can only be maximised when the interaction of soil biota with environmental factors, including temperature, water content, and energy source is understood. There is currently a fundamental knowledge gap in the interpretation and linking of various proposed biological, chemical and physical indicators. Measurement protocols and indexing techniques are needed for easy identification of the soil properties, processes and the effects of human management practices over time.

Soil quality assessment and interpretation tools must be sensitive and responsive to the various soil properties and processes that respond to changes in soil and crop management practices and land use decisions. They also need to account for differences in inherent soil conditions among various physiographic regions and their response, both positive and negative, to management practices. They should help determine appropriate land uses and input needs and help land owners and operators to select or develop more environmentally-sound management practices, while providing the food, feed and fibre needed to satisfy increasing human needs.

The assessment of the health of soils, through the identification of key soil properties, which can serve as indicators of soil health, has become a major issue for land managers and the food and agricultural sector through the world. For example, FAO has recently been identified as the executing agency for conducting, in close cooperation with multiple partners, the GEF/UNEP Land Degradation Assessment in Drylands (LADA).

Soil biota and its functions should be a key component of such assessments. In particular, there is a need to determine short- and long-term effects of agricultural management practices on soil biological community populations, biodiversity, functioning and resilience. Relating soil quality/ health to productivity, in terms of crop yield and profitability, and environmental effects from drainage, leaching, runoff and erosion is essential in order to evaluate the sustainability of various land management strategies.

There is recent progress in realising that soil health, by its broadest definition, is inseparable from issues of sustainability. The challenge ahead is to develop holistic approaches for assessing soil quality and health that are useful to producers, specialists and policy makers in identifying agricultural and land use management systems that are profitable and will sustain finite soil resources for future generations. Benefits of paying more attention to soil health and its assessment include its potential use in: the evaluation of land-use policy and of practices that degrade or improve the soil resource; and in the identification of critical landscapes or management systems and of gaps in our knowledge base and understanding of sustainable management.

b) Managing Interactions among land management, soil biodiversity and agricultural production

Land use and the type of farming system impacts upon soil life, while soil management controls and manipulates the organisms responsible for nutrient cycling, crop diseases and pest damage through its effects on soil physical and chemical conditions, biological habitat, food sources and plant-host interactions. Biotic processes impact on long-term productivity, soil fertility, soil aggregation, erosion and other indicators of soil quality. In turn, the soil biota and their interactions play a part in the success of any management decision.

For example, intensive cultivation coupled with mono-cropping practices may detrimentally affect the functioning of the soil biota leading to loss of plant nutrients and soil aggregate structure and resulting in soil degradation, environmental pollution and declining crop yields. On the other hand, minimum tillage practices and better crop cover, coupled with a more diverse cropping regime, may promote the more effective functioning of soil biota, resulting in improved soil structure and nutrient and water management and hence crop productivity.

Case study of Biological nitrogen fixation: The natural process of biological nitrogen fixation (BNF) constitutes an important source of nitrogen for crop growth and protein production in many soils and ecosystems. It therefore provides a major alternative to the use of commercial nitrogen fertiliser in agriculture. It has recently been estimated that global terrestrial BNF ranges between 100 and 290 million tons of nitrogen per year of which 40-48 million tons N per year is estimated to be biologically fixed in agricultural crops and fields. In comparison, 83 million tons per year are currently fixed industrially for the production of fertiliser.

Biologically fixed N_2, either asymbiotic, associative or symbiotic, is considered a renewable resource, which should constitute an integral part of sustainable agro-ecosystems globally. The contribution of legume N fixation to the N-economy of any ecosystem is mediated by: the efficiency of the N_2 fixing system; the contribution of BNF to the soil N pool; and the total amount of N_2 fixed that actually is recycled by human practices and animal manure into the system.

Several opportunities to enhance BNF inputs are available across different agro-ecosystems and socio-economic conditions, inter alia: through altering the number of effective symbiotic or associated organisms in the system (inoculation); screening and selection of the appropriate legume crop (selecting high BNF species well adapted to environmental conditions); and management practices that enhance N_2 fixation and recycling of net N_2 inputs into the cropping system (rotation, green manure application, no-tillage, strategic use of legumes, etc.).

Gaps and needs. The complex relationships between soil biota, ecosystem functioning and land management practices must be well understood in order to develop guidelines for agriculture that will optimise resilience and sustainability of the ecosystem. A better understanding of the ecology of beneficial and harmful organisms is needed to utilise and control their expression in agricultural systems. An understanding of soil biota and their ecology must be developed, so that the ecological and biological effects of resident soil populations can be used to reduce inputs of non-renewable resources while still increasing productivity needed to meet food, feed and fibre demand.

Opportunities/areas for action. There is a need to enhance scientific and farm knowledge of soil biota-manipulation and ecosystem interactions to obtain better understanding of the processes they control and, thereby, to influence plant growth, soil biotic functions and soil productivity. This includes:

— development of fundamental understanding of the ecological characteristics and processes of the soil and root biology to predict accurately root, seed, soil and soil biota interactions;

— identification of fertility, cultural, spatial and temporal factors affecting these interactions;

— development of effective strategies to manage soil biota as an integrated aspect of soil and land resources management;

— development of improved methods to identify and characterise soil biota populations and their activities for farmer level in order to help in the interpretation of interactions between farmers' practices, soil function and agricultural production.

c) Soil biodiversity and biological management of pests

The rate and extent of build-up or maintenance of indigenous or introduced pathogens or pests depend on many environmental and cultural factors, including residues, organic matter and cover crop issues, plant stress, soil tillage, poor irrigation management and fertilisation practices and crop genetics. Intensive cropping, monocropping and the over-use of agrochemicals often increases the build up of soil-borne pathogens (disease-carrying organisms), pests and weeds. This is also reflected following conversion to reduced or no-tillage practices, when carefully controlled herbicide use and prudent pest management practices may be required in the initial years until an ecological balance is restored and the natural biocontrol mechanisms become reestablished. Under no-tillage it has been reported that pathogens, pests and weeds are not necessarily greater but may differ from those prevalent under tilled systems; with appropriate management under no-tillage the equilibrium tends to favour beneficial organisms.

Soil biota can influence the growth of some organisms including larger life forms such as certain insects, crop plants and weeds, both positively and negatively. In some cases, deterioration of soil productivity stems from changes in soil biotic communities, reducing their capacity to suppress root pathogens and pests by biological means. Pathogens and pests unchecked by ecological competition can achieve populations that are devastating to agriculture and pose serious threats to economic sustainability. The nature of the pest outbreak, whether bacterial, fungal, viral, nematode, insect or weed, indicates the kind of management strategies needed to restrict or eliminate its activities. The strategies available to farmers are cultural (cropping practice), chemical and biological; however, not all strategies are feasible for every cropping system. Ecologically-oriented pest management within a viable, integrated systems' approach is gaining popularity. Management of the edaphic (soil-based) phase of the life cycle needs to be explored to develop additional biological pest management options.

Case study of alternatives to methyl bromide in managing pests: Under the Montreal Protocol of 1991, methyl bromide was defined as a chemical that contributes to depletion of the Earth's ozone layer; and it was internationally agreed that consumption of this product will be frozen in developing countries in 2002. Farmers who are dependent on methyl bromide for suppressing soil-borne pests and diseases are having to shift towards more environmentally sustainable agricultural practices. Alternatives to the use of methyl bromide have been investigated and biofumigation is one such example, that uses the Brassica family (i.e. broccoli, cabbage, cauliflower and rape) for producing toxic compounds.

Gaps and needs. Use of soil biota in pest management could increase crop efficiency, decrease the need for tillage and decrease the use of synthetic chemical pesticides. Often individual pathogens have been studied in isolation, which limits knowledge of activities in situ with the whole biotic community. A greater awareness of the full range of the soil biota community and its impact on its own soil community dynamics, plant growth and chemical-plant interactions are critical. Integrative approaches have the potential to be used to manage the production system and natural soil organism-plant interactions for pest suppression, either from adding beneficial organisms that can suppress the pests or managing or increasing such organisms that are resident in the soil. Further study is required, so that the ecological and biological effects of the resident soil organism population on pest growth can be used effectively in pest management strategies. Moreover, the use of soil biotic dynamics and integrated approaches to managing soil-borne pathogens or pests may also require additional soil management practices.

Soil micro-fauna play an important role in suppression of plant pathogens and represent a significant biological control potential. There are opportunities to develop effective and economically feasible disease and pest control strategies that reduce pathogens and pests through the introduction of antagonists or by managing resident soil biota to increase their activity. Efforts to manipulate and exploit the friendly fauna populations for crop benefit must be compatible with microbial symbionts, and other plant-growth promoting rhizosphere organisms, and with fungi and bacteria that are being promoted for biological control of diseases. This is clearly an area with great opportunities for further research.

d) Bioremediation efforts

The goal of bioremediation efforts is to reduce the potential toxicity of chemical contaminants in the field by using micro-organisms, plants and animals to transform, metabolise, remove or immobilise toxicants. There is already a significant knowledge-base about many pathways for organic degradation, and several important contaminant degradation mechanisms are under detailed investigation.

Different types of organisms can be bioremediation agents, for example, micro-organisms (primarily bacteria and fungi) are nature's original recyclers. Their capability to transform natural and synthetic chemicals into sources of energy and raw materials for their own growth highlights their value as cheaper and more environmentally-benign alternatives to chemical or physical remediation processes.

Plant roots can also indirectly stimulate microbial degradation of contaminants in the rhizosphere. The intrinsic ability of certain plants for uptake, translocation, transformation and detoxification of contaminants also offers a newly recognised resource that can be exploited. Research continues to discover and verify the bioremediation potential and unique properties of many organisms.

Case study of Bioremediation: These techniques are used to remove environmental pollutants from sites where they have been released or more often to reduce their concentrations to levels considered acceptable to site owners and/or regulatory agencies. Many bioremediation techniques exist to treat in situ soil contaminants and a number of organisms have been involved, particularly bacteria – such as Achromobacter, Acinetobacter, Alcaligenes, Bacillus, Nocardia, Pseudomonas – and fungi such as Trichoderma, Rhodotorula, Mirtirella, Aspergillus. The rate at which microbial communities adapt their metabolism to toxic compounds is crucial in bioremediation.

A recent addition to the growing list of bacteria that can sequester or reduce metals is Geobacter metallireducens, which removes uranium, a radioactive waste, from drainage waters in mining operations and from contaminated groundwater. The concept of phytoremediation—the use of plants for abatement and containment of pollution—is developing as an acceptable management technique. This concept is also being applied in other environments, such as riparian zones and filter strips.

Gaps and needs. A tiny fraction of the soil microbial diversity of the Earth has been identified, and an even smaller fraction has been examined for its biodegradation potential. Understanding of biochemical and transformations of contaminants in soil has advanced in recent years. However, knowledge of the specific pathways for degradation/ detoxification and of the role of specific organisms and communities is limited. Biological approaches on the molecular level can clarify the expression and regulation of xenobiotic (contaminant) degradation and help provide methods to develop plants and micro-organisms with enhanced detoxification ability. This knowledge is essential for understanding the ability of soil to maintain a biological buffering barrier for pollution and in the design of systems to decontaminate soil and water.

Despite the successful contributions of existing knowledge, the understanding of biotransformation and biodegradation pathways and mechanisms in the field is incomplete. Opportunities are wide for further research (for example of microbial physiology and ecology, enzymology, biochemistry and plant-micro-organism interactions) and technology applications. Opportunities exist for the development of knowledge and techniques that will minimise the impact of agrochemicals and other xenobiotics in the environment and of approaches to promote the degradation of xenobiotics in soils. Improved methods and decision-making tool are needed for soils that require remediation with a view to improving soil productivity, protecting human health and preventing environmental degradation.

e) The ecological impact of agricultural biotechnology.

Agricultural biotechnology, if appropriately integrated with other technologies, offers opportunities for developing more productive and sustainable systems, for example the development of plant varieties and animal races for overcoming specific constraints such as soil and climatic limitations and specific pests and diseases. Biotechnology includes a wide array of techniques and applications, from natural fermentation processes and cell culture to genetic engineering, protein engineering and DNA amplification. Transgenic materials provide greatly increased opportunities but also potentially significant risks of affecting soil biodiversity and the ecosystem at all levels, for example, through upsetting the delicately balanced and complex food web.

There is a need to assist national and local governments in the formulation and application of policies that ensure the proper ownership and receiving of benefits deriving from the use of soil biodiversity, in particular the technologies and products that derive from the manipulation and extraction of particular components of the soil biota (especially micro-organisms and their products). Taking into account issues of bioprospecting, traditional knowledge and farmers' rights, this raises the important consideration of finding ways in which soil biodiversity and associated knowledge systems, can be managed for the benefit of farmers and rural communities and to ensure that legal and international property rights regimes support this aim.

An example of the beneficial use of biotechnology in the management of soil biodiversity is in the development of improved microbial inoculants. Effective wild-type strains are isolated from the environment for use as microbial inoculants in agriculture and recombinant DNA technology (i.e. genetic engineering) may be used to further improve microbial strains.

Microbial characteristics that are being targeted for improved inoculant performance include: the survival ability of the inoculated strain, as in the case of strains that are better adjusted to soil constraints such as salinity, acidity or aridity; competitive nodulation of legume roots, as in the case of symbiotic nitrogen-fixing Rhizobium bacteria); and interactions with beneficial micro-organisms, for example compatibility with mycorrhizal fungi, and interactions with detrimental micro-organisms, for example for the inhibition of plant pathogens in the rhizosphere.

On the other hand, the area under transgenic crops is rapidly expanding and yet it is not well known what might be the long term effects on the ecosystem of potentially higher herbicide applications or the indirect effects of transgenic plant root exudates (secretions). The decomposition of modified genetic material from plant remains in the soil could seriously affect the balance of soil micro-organisms and be an ideal medium for horizontal gene transfer. Incorporation into plants of genes from Bacillus thuringiensis (Bt) toxins that code for the production of insecticidal toxins can be incorporated into the soil through leaf materials, when farmers incorporate crop residues after harvest.

Toxins may persist for 2-3 months, resisting degradation by binding to clay and humic acid soil particles while maintaining toxin activity. Such active Bt toxins that end up and accumulate in the soil and water from transgenic leaf litter may have negative impacts on soil and aquatic invertebrates and nutrient cycling processes. Increased and frequent use of glyphosphate applications has produced changes in the microbial composition of soil in the field associated with "Roundup Ready" soybean production. The use of glyphosphate-resistant soybean changes the dominance of fungi versus bacteria in the soil, altering nutrient cycling processes, nutrient retention ability and the ability of the soil to suppress disease. There has been little attention to monitor and improve understanding of the effects of transgenic crop plants, such as herbicide resistant soya beans or cereals, on soil biodiversity and their functions.

Perturbations have been recorded by several authors with the introduction in the soil of genetically modified micro-organisms (such as Pseudomonas fluorescens), including displacement of indigenous populations, suppression of fungal populations, reduced protozoa populations, altered soil enzymatic activity, and increased carbon turnover. Circumstantial evidence that genetic exchanges between strains of Rhizobia occur in a field environment has been provided by population studies. However, information on the time scale and on the conditions in which these exchanges take place, is still missing. More research on the consequences of the release of novel organisms in the rhizosphere before they can be safely utilised is necessary.

Genetically modified organisms (GMOs) need to be adequately assessed for their environmental or human health effects before they are released into the environment. However, it is very difficult to predict how GMOs will behave once in the agricultural ecosystem. Today, results show that soil organisms are extremely sensitive to the use of engineered plants, and the effects are unpredictable.

The impact of modern biotechnology on the environment and on human and animal health needs careful assessment on a case by case basis and through applying, in each situation, the precautionary approach, as adopted by the CBD. Attention is drawn to the need to consider how to implement the precautionary approach effectively and thereby address the concerns over risks and potential benefits of GMOs. International bodies such as FAO, UNEP, UNESCO and the CBD process, in particular the Biosafety Protocol, may provide guidance and assistance to countries on this matter. However, final decisions on the use of biotechnology remain a national responsibility.

International Framework on Soil biodiversity Conservation

The current set of international environmental conventions have been developed on the basis of the global policy statement—Agenda 21 Plan of Action – that was adopted at the UN Conference on Environment and Development in Rio 1992. This "Earth Summit" called for countries to incorporate environmental considerations into their development plans and build national strategies for sustainable development. At the United Nations General Assembly's special session in 1997 – "Rio plus five"—countries agreed to have such national strategies in place by 2002, which should be the product of extensive consultation with the stakeholders concerned. Countries are being assisted by donors to develop and implement these national strategies. A "Rio plus ten" summit will take place in Johannesburg in 2002 to assess progress achieved since 1992. The national strategies for sustainable development provide a useful framework for addressing issues of soil biodiversity management and conservation as part of an integrated approach.

The UN Convention to Combat Desertification (CCD) aims to address land degradation and drought in dryland areas, with the aim of improving living conditions. The text of this Convention binds signatory governments to promote long-term integrated strategies to improve the productivity of land, rehabilitate degraded areas, and conserve and manage land and water resources in a sustainable fashion, in particular at community level. National Action Programmes to address land degradation are being drawn up by a large number of countries through a consultative process, for which donor support is being sought. Soil biological management, including the conservation and sustainable use of soil biodiversity and its functions, should be an integral part of such plans.

The UN Framework Convention on Climate Change (FCCC) aims to achieve stabilisation of greenhouse gas concentrations in the atmosphere at a level that will prevent dangerous interference with the climate system.

Such a level should be achieved within a time frame sufficient to allow ecosystems to adapt naturally to climate change, to ensure that food production is not threatened and to enable economic development to proceed in a sustainable manner. The Kyoto protocol, which aims at a reduction of carbon dioxide emissions, was drafted in 1997, and awaits ratification. There are various links between climate change and soil management especially in regard to carbon sequestration (the storage or fixation of Carbon in soil organic matter and in plant biomass) and greenhouse gas emissions (GHG).

The most important greenhouse gases are carbon dioxide (CO_2), methane (CH_4) and nitrous oxide (N_2O). Photosynthesis in plants leads to carbon fixation and CO_2. Decomposition and burning of biomass, however, releases CO_2 back to the atmosphere. Methane is produced in wetlands and rice fields, and by ruminant animals. Soils also emit N_2O as a result of microbial processes. At a global level, the mining, manufacture and transport of mineral fertilisers contribute to CO_2 and N_2O emissions. Thus, changes to soil fertility management by incorporating or enhancing biological management of soil fertility could have significant implications for climate change.

Agriculture provides a major share of national income and export earnings in many developing countries, while ensuring food security, income and employment to a large proportion of the population. Farmers, governments and scientists are increasingly aware that declining soil fertility is becoming a major concern worldwide with social, food security and environmental implications. As a result, controlling erosion and improving the management of soil fertility have become a major issue on the development policy agenda. In this regard, the Soil

Fertility Initiative (SFI) for Sub-Saharan Africa was launched as part of the Rome Declaration on World Food Security in 1996, among key collaborating organisations, including the World Bank, FAO, ICRAF, IFDC, IFA, IFPRI and USAID. This interactive process aimed at increasing synergies and catalysing comprehensive strategies and actions at country level to enhance soil fertility restoration and management and prevent further nutrient mining.

The development of Soil Management Action Plans has been promoted in over 20 countries through participatory review and prioritisation processes. In Burkina Faso and Ghana, for example, such plans have been developed and approved by the government. In other countries, certain priority actions are being addressed through investment and technical assistance programmes and with donor support. A multitude of programmes in the agricultural and land sectors are supporting improved soil and land resources management and provide great scope for expanding attention to the conservation and sustainable use of soil biodiversity and the important functions of soil organisms.

FAOs support to Member countries could be further mobilised to integrate soil biodiversity management through, inter alia: projects to improve capacities and tools and farmer-led learning approaches for soil productivity improvement and conservation agriculture, initiated through its Technical Cooperation Programme, the Special Program on Food Security and work on sustainable livelihoods; projects to mitigate land degradation and promote integrated watershed management and production systems; and, the CBD/FAO joint Programme of Work on Biodiversity for Food and Agriculture.

More specifically, the FAO-Netherlands Partnership Programme (FNPP) is a two-year programme that is supporting work by FAO towards the conservation and use of agricultural biodiversity within sustainable ecosystems and its contribution to global food security. One of the four main areas of attention is on improving understanding and implementation of the ecosystem approach, including adaptive management and best practices. In this regard, the sub-component on soil biodiversity aims to help catalyse more applied work in the agricultural and land sectors with the support of scientific institutes that are currently focusing their research on certain categories and functions of soil biodiversity and on specific technologies. There are three main aims and axes of cooperation:

— Sharing of knowledge and information on the roles of diverse soil organisms in providing key goods and services and the impacts of existing and new agricultural technologies and management practices, with a view to developing guidance for agricultural and environmental-CBD fora;

— Collaboration among relevant programmes, networks and national and inter-national bodies to identify and promote improved soil biological management practices for different conditions and their integration into ongoing land management and soil productivity efforts; and,

— Establishing partnerships among farmers/land resource users and researchers/ development programmes to monitor and assess different practices and prepare case studies and to integrate soil biodiversity issues into documentation and training materials.

2

Strategy for Food Security

Over the past three decades, the world has made remarkable progress in increasing food production and reducing food insecurity. But progress slowed considerably during the 1990s, and achieving sustainable food security for all remains an elusive goal. Over the course of the 1970s and 1980s, according to the Food and Agriculture Organisation of the United Nations (FAO), the food-insecure proportion of the developing world's population fell steeply, from 37 percent to 20 percent, whereas it declined only slightly, from 20 percent to 17 percent, during the 1990s. Likewise, the number of food-insecure people dropped by 14 percent, from 959 million to 824 million, between 1970 and 1990, meaning an average annual decline of 7 million people.

During the 1990s, the number of foodinsecure people decreased by just 1 percent, or less than 1 million people per year. If China is excluded, the number actually increased by more than 40 million people. In recognition of the need to reinvigorate the international community's focus on these critical issues, the member states of the United Nations in 2000 agreed on eight Millennium Development Goals (MDGs)-specific, measurable targets to be met by 2015 that will make definite improvements in the lives of the world's poor and hungry people. The MDGs represent a significant revision of the fundamental paradigm of development thinking and have become an important focal point for food security and development policy.

Achieving the MDG of cutting hunger in half by 2015 will be especially challenging. One thing is clear: this goal will not be achieved through business as usual. Research is needed on the barriers to accelerating this sluggish progress of eliminating hunger in a food-rich world. The political and economic context has changed. Rapid changes are taking place in the structure and authority of governments, the global economy, the structure of the farming sector, and global and local food industries and retail businesses.

The fundamental roles of governments are changing. In the "layered societies" of the past, maintaining a good working relationship with key central government ministries was often considered sufficient to get food policy research translated into improved policies. In today's "network societies," government authority is more diffuse and other actors are involved, including local governments, business and industry, nongovernmental organisations (NGOs), and other parts of civil society. In numerous instances, especially in Sub-Saharan Africa, states have failed and protracted civil wars have proliferated.

In many countries, NGOs have rapidly expanded their role in food security and natural resource management debate and action. Market liberalisation and globalisation are powerful forces transforming the global economy. Yet the opening of economies in both developed and developing countries poses difficult challenges for developing-country food security, agriculture, and natural resource management. Many obstacles may prevent low-income countries from capturing the benefits of globalisation. Risks include the short-term inability of many developing-country industries to compete, the potential destabilising effects of short-term capital flows, increased exposure to price risks, and worsening inequality within and between countries. Public-sector leadership is needed to facilitate privatisation and guide the transformation of agriculture in a pro-poor direction.

The nature of the farming sector is changing in many developing countries. Smallscale family farms are under pressure, threatened by biased investments that encourage larger-scale production. Often, poor people lack alternative income sources and migrate, so poverty and food insecurity-while still predominantly rural-are steadily urbanising. These issues are further complicated by population growth, the aging of the farm population, increasing demands on women's time at home and on the farm, the decreasing cost of capital relative to labor, and the depletion of asset bases resulting from man-made and natural disasters, which are affecting growing numbers of people.

At the same time, global and national food systems are increasingly driven by consumer interests, changing consumption patterns, and food quality and safety concerns. Food processing and retail industries are responding, profoundly affecting production, markets, trade, diets, and public policy. At the global level, transnational corporations and broad NGO coalitions are becoming increasingly prominent and influential in policy debates.

Technological innovation may bypass poor people. New technological developments related to food, agriculture, nutrition, biotechnology, energy, and information and communications offer great opportunities to improve poor people's food security. Within the Consultative Group on International Agricultural Research (CGIAR) system, the policy research challenge is to identify and target high-priority biological research and development (R&D) to solve critical problems facing small farmers and poor consumers.

Satellites and geographic information systems have great potential to help researchers and policymakers collect data and analyse spatial issues related to the production and distribution of agricultural commodities, natural resource management, and poverty eradication. Solar panels, cell phones, and other new communications and energy technologies could also provide significant benefits to poor people in developing countries. Research is needed to identify the policies that will make these technologies accessible to food-insecure people. Much of the latest scientific research in the area of food and agriculture is marketdriven and hence focused on meeting the demands of well-off people in rich countries.

The research environment is changing, with strengthened national systems in some developing countries and weakened institutions in others. Private food and agriculture-related companies have rapidly expanded R&D activities, and in principle there should be tremendous opportunities for public-private partnerships, especially in developing countries.

At a global level, few meaningful results have materialised from such partnerships so far, probably because research is taking place in a business climate driven by concentration through mergers and short-term shareholder expectations of gain, with no consideration for markets in which poor consumers and small farmers operate. Moreover, the products and processes of research are increasingly subject to intellectual property rights protection in an uncertain legal environment. To reach poor farmers and consumers, public agricultural research must continue to play a key role in developing countries.

Health and nutrition crises go unaddressed. Broader nutrition-related health problems remain insufficiently addressed. HIV/AIDS, tuberculosis, malaria, micronutrient deficiencies, and chronic diseases are all compromising food and nutrition security in many developing countries. HIV/AIDS affected over 39 million people in 2004. AIDS has a direct impact on people's ability to acquire enough nutritious food to lead active, healthy lives. It turns millions of children into orphans, disrupts social bonds and the transmission of agricultural knowledge from one generation to the next, weakens informal property rights, and undermines people's ability to engage in collective action.

It reduces availability of farm labor and can force people to devote less time to farming and more time to patient care. Tuberculosis kills 30 percent of AIDS victims in Africa and Asia, while AIDS accelerates the progression of tuberculosis by up to a hundredfold. African preschoolers account for 90 percent of malaria deaths. Because malaria often strikes during harvest time, it threatens food security. Micronutrient malnutrition, often called "hidden hunger," afflicts more than 2 billion people, with a devastating impact on health and productivity.

All countries of the African region have committed themselves to the achievement of development goals which were set at a series of UN conferences during the 1990s. Most of these goals have also been formally adopted by the donor community as represented by the United Nations, the World Bank, the International

Monetary Fund (IMF) and the Organisation for Economic Co-operation and Development (OECD). The goals are as relevant to the Horn of Africa region as they are to the rest of the world. It is logical, therefore, to adopt them as the targets at which a joint government-UN partnership will aim.

Central to the objective of improving food security is the commitment made at the World Food Summit to reduce by half the number of undernourished people worldwide by 2015. In regional terms, this implies a goal of cutting the number of chronically undernourished people from around 70 million to 35 million by 2015. A first step towards achieving this goal must be to ensure that the threat of famine will never be allowed to recur. Through concerted commitment and action, this target may even be exceeded.

Other international goals for 2015 that are of relevance to the region and to the elimination of food insecurity, and for which region- and country-specific targets and indicators that can be monitored would be developed, include:

— reduction in the proportion of people living in extreme poverty by at least half;

— universal primary education;

— reduction in infant and child mortality by two-thirds of the 1990 level;

— reduction in maternal mortality by three-quarters of the 1990 level;

— access through primary health care systems to reproductive health services for all individuals of appropriate ages;

— reduction by half of the proportion of people who lack sustainable access to affordable and safe water;

— reversal of current trends in environmental degradation and compliance with the international conventions on conservation of biodiversity, climatic change and desertification control.

Right to Food—A Fundamental Human Rights Framework

The right to food is among the fundamental human rights enshrined in the UN Universal Declaration of Human Rights, to which all nations in the Horn of Africa have subscribed. The Universal Declaration and subsequent covenants provide a useful starting point for any long-term strategy through which the concerned governments and the UN system can jointly address food insecurity and deprivation in the region. These rights are enshrined in international law and have been reiterated in the declarations of the 1990s summits and international conferences. The regular recurrence of famine in the region is evidence that rights have been infringed, whether wantonly or inadvertently.

The formulation of a long-term strategy to address the problem of food insecurity in the Horn of Africa is something that must be, first and foremost, the responsibility of the governments of the region. When respect for the fundamental human right to freedom from hunger and deprivation is adopted as a central goal of development, and its implications are acknowledged, it is possible to define a set of broad strategic principles which can guide the way ahead. Full respect for human rights, and particularly the right to food, can create an enabling environment for achieving improved food security and can become a powerful tool in mobilising the required resources. It requires that governments:

— desist from actions that undermine the food security and well-being of any part of the population. In particular, governments must desist from using food as a weapon in times of conflict. When this fundamental obligation has been ignored, the consequences in terms of human suffering have been terrible;

— actively pursue policies that address the problem of poverty and food insecurity, if necessary with external assistance when resource requirements exceed those that can be mobilised domestically. The obligation to provide basic security, thus enabling people to exercise their rights to acquire food, is fundamental. Particularly in the pastoral and agropastoral areas, which are especially prone to the effects of drought, communal and individual rights have been regularly infringed or ignored;

— provide an enabling environment in which all parts of the population can thrive. This means encouraging all actors—the private sector, non-governmental organisations (NGOs), community-based organisations (CBOs) and individuals—to pursue activities that will ensure inclusive food security and prosperity.

Within this framework, governments and their partners can address the critical constraints to improved livelihoods by:

— *Allocating greater resources*: Human and financial, national and external resources are needed to combat both the immediate and long-term problems of food insecurity and poverty.

— *Strengthening entitlements*: The entitlements of all food-insecure people can be enhanced through helping them to broaden their livelihoods in a sustainable way, and by building on and strengthening existing coping mechanisms.

— *Increasing household income and output*: All viable technological opportunities that are appropriate and sustainable should be exploited.

— *Enhancing knowledge and information systems*: Households and communities, especially in remote and marginal areas, must have access to the knowledge they need to be able to survive and prosper, and governments and the international community must have the information to be able to react quickly and efficiently when emergency situations loom and to formulate appropriate mitigation and development policies.

— *Strengthening the overall environment*: The framework for enhanced economic activity and reduced risk, especially in remote and marginal areas, must be improved through better governance, including decentralisation of administration and resources, avoidance of conflict, provision of physical infrastructure that will enhance access to markets, improvement of social services to raise the overall health and educational status of vulnerable people and a policy framework that encourages increased production as well as growth in domestic, regional and international trade.

Ensuring Food Security

The rights-based framework recognises that some people will not benefit from improved policies and will remain at risk of food insecurity and that existing coping mechanisms may fail, especially in the face of drought or other hazards. For those who are unable to take advantage of the opportunities for improving their livelihoods that become available, governments are obliged to ensure that social safety nets are in place in order to prevent undue deprivation without compromising human dignity.

However, the history of direct intervention in the provision of food by governments through price controls and rationing should serve as a clear warning of the need for the careful planning and direction of any actions in the future. Government capacity to take full responsibility for poverty eradication and food security is often highly limited and, particularly in the Horn, external agencies (including both relief and developmental organisations) must continue to play a central role in supporting governments' efforts to fulfil their obligations.

Role of Government

Recent episodes of famine or acute food insecurity in many parts of the region show that the previous efforts of governments and their development partners have been both inadequate in scale and persistence and, frequently, misdirected. Some well-intentioned interventions have had the unintended effect of increasing, rather than reducing, vulnerability to drought and other adverse events. For example, the injudicious development of water for livestock in some pastoral areas has led to the expansion of herds beyond the capacity of the range, resulting in environmental degradation and reduced resilience in time of drought.

The strategy must be refocused so that the driving force comes from communities and local institutions, who are ultimately responsible for identifying, formulating and implementing programmes and projects. The strategy must provide a greater role for civil society organisations (CSOs), including farmers' unions and CBOs, NGOs and the private sector, and explore alternative investment channels that promote rather than undermine self-reliance. Within such a strategy, while there is a place for certain major investments in large-scale infrastructure items (such as improved ports and highways), the bulk of investments should be in response to local preferences and demands, and should avoid top-down "technical solutions". Appropriate financial mechanisms and participatory design processes must be used, devolving the responsibility for choices to the local level. The details of national action and local investment plans must be generated from the countries and localities concerned.

The beneficiaries of past government and donor-financed programmes have often been limited in number and have seldom been the most needy people. Typically, the poorest people, especially those living in remote areas, have little voice in the design of programmes, and therefore do not reap the benefits. International investment in poverty alleviation should be directed to the least privileged elements of society. Even if programmes do not have a specific poverty alleviation or food security focus, due attention must be paid to distribution of the costs and benefits of investments and to the implications of programmes for household and community claims on resources.

There is a need to shift the bias towards the people living in remote, fragile and highly famine-prone highland areas and arid and semi-arid lowland areas throughout the Horn. This shift must include a focus on pastoral and agropastoral people, who have been largely neglected by government services and investment projects and who continue to have very little influence on national policies. The needs of marginalised groups in urban areas, such as street children, should also be considered. Most importantly, mechanisms must be created, especially at the community level, to ensure that excluded people are able to participate in the design and implementation of programmes and share in their benefits.

The most important response by the international community to the realisation that pro-growth policy reforms have not reached the poorest people and the remotest parts of countries has been to put poverty reduction at the top of the development agenda. This is reflected in the focus of bilateral donor programmes and in the Poverty Reduction Strategies (PRS) being promoted by the Bretton Woods institutions. The World Bank's and IMF's decision to make client governments' adoption of pro-poor, poverty reduction policies the principal criterion for the approval of assistance marks a fundamental change in the way in which the international community operates.

Women's Food Security

Women are critical to food security in the Horn of Africa. They are the primary agricultural producers, the preparers of food and the carers of the family. However, they are more vulnerable than men, for a variety of reasons. They are typically illiterate and have neither basic education nor appropriate technical skills. They invariably lack access in their own right to productive assets such as arable land and inputs for production, and undertake the majority of agricultural tasks with just simple tools and by working long hours. Women suffer discrimination because they have no recognised independent status as farmers, and their contribution is considered as secondary, both within the family and in society. Their livelihood strategies are based on successfully managing the natural resources to which they have access, but these are often the hardest hit when disasters occur.

Customary traditional values and laws limit women's opportunities to participate in local decision-making processes and restrict their access to credit, research and the use of improved technologies. Despite their important contributions to the household economy, food security and sustainable family livelihoods, women are excluded from decision-making at the household, community and national levels because of these restrictions. Women's local knowledge of the environment and community natural resources, together with their social networks, offer important avenues for both disaster mitigation and development activities which should not be neglected.

Both women and men have gender-defined roles and capacities that must be taken into consideration when planning and implementing programmes intended to reduce vulnerability and develop community self-reliance. Women's coping strategies and skills should be reinforced and utilised, not only for disaster mitigation but also for eliminating long-term food insecurity. At the same time, they need to be protected from extra work loads and their involvement in decision-making processes should be encouraged through broadening female participation in local organisations.

Food Security for Sustainable Livelihoods

Offering people the tools to attain prosperity entails considering, not just how to increase crop and livestock production, but also how to open up alternative sources of income. In the marginal areas it also means building on people's own ways of dealing with risk. Where crop- and/or livestock-based production systems can provide neither a sufficient nor a reliable livelihood, diversification out of agriculture may be the only sustainable and viable option. It is almost impossible for external agents to identify activities that will support sustainable livelihoods. Priorities must be set by the communities themselves, as they know more than any outsider about the constraints they face and the opportunities available.

The basic philosophy behind the strategy recognises that there is no single solution to the underlying problems of deprivation in the Horn of Africa and that synergies among different elements of the strategy must be exploited. For example, efforts to introduce more efficient agricultural technologies are likely to fail if people are not sufficiently educated to understand their implications or if the transport infrastructure is too weak to permit efficient marketing and input supplies. Similarly, nutrition interventions will have little impact if mothers do not have a basic understanding of child care, have no access to basic preventive health services and have no clean water supply. The Sustainable Livelihoods Approach (SLA) emphasises full participation at the local level, and recognises the need for policies that are informed by local people's knowledge and insights.

Long-term Strategy

While it is certainly possible—and necessary—to avoid recurrence of famine, it will take many years to achieve better livelihoods under the very difficult resource-poor conditions in which the region's most vulnerable people live. This is a process that involves fundamental social transformation, ultimately leading to a better equilibrium between population and natural resources. Improved nutrition and sanitation will reduce disease and premature mortality and enable people to gain more from improved education, which in turn will lead to a progressive slowing of population growth rates. Better health and education will enhance creativity, opening up more opportunities for economic diversification and raising people's competitiveness in labour markets. It will also enable stronger local institutions to emerge.

These are long-term processes that cannot be externally driven. They can, however, be facilitated and nurtured by the careful and sensitive provision of external financial and technical support in response to local needs and offered in a spirit of mutual confidence. Critical to the success of any interventions will be an acceptance by partners, local and external, of the need for a long-term and reliable engagement that spans many more years than the typical development project.

3

World Food Security Trends

World food production has grown faster than population, over the past three decades. Per caput food production is today about 18 percent above that of 30 years ago. Food availabilities for direct human consumption are equivalent to some 2,700 calories per person per day, up from 2,300 calories 30 years ago. At the one extreme, in Western Europe per caput food availabilities stand at some 3, 500 calories and in North America at some 3,600. At the other extreme, average per caput food availabilities are only 2,170 calories in sub-Saharan Africa.

Despite the considerable progress achieved in increasing per caput food supplies, more than 800 million people were undernourished in the early 1990s. Millions more suffer debilitating diseases related to micronutrient deficiencies and to contaminated food and water. Every day, one out of five people in the developing world cannot get enough food to meet his/her daily needs; in sub-Saharan Africa, two out of five people do not have adequate food. The Near East and Latin America and Caribbean regions had the lowest percentage of undernourished. The largest numbers, though declining, are to be found in Asia but those in sub-Saharan Africa have been increasing rapidly, both in total and as a proportion of the region's population.

In addition to the chronically undernourished, civil strife and wars have adversely affected millions of people. Although food assistance is provided to ease their plight, the amount provided for each individual is too often insufficient for good health. The sharp reduction in food aid availability over the past three years has reduced our capacity to face crises. To bring each undernourished person to his or her respective energy requirement level (2 200 calories/day on average) would require an additional average 570 calories/day. This is obviously an underestimate of any realistic scheme to eliminate undernourishment exclusively, but the resulting order of magnitude is telling.

World food consumption in 1990-92 would need to be higher by about 3 percent. In more concrete terms, given that cereals represent around 60 percent of the calorie supply of the population of the developing countries, this increase would represent about 30 million tons of grains (to be compared with about 9-12 million tons of food aid in recent years). This food gap estimate varies widely between regions, ranging from negligible in the western industrialised countries to about 5 percent in the low-income food-deficit countries, 10 percent in Africa and close to 5 percent in the developing countries as a whole.

The prospects for the future, as they emerge from *FAO's World Agriculture: Towards 2010 (AT2010)* study (1995), indicate that trends towards increasing per caput food supplies in most developing countries will continue. For the developing countries as a whole, average per caput food supplies are expected to reach 2,730 calories in 2010—a substantial increase from 2,520 in the year 1990-92.

Despite such progress, the developing countries are projected to continue to have some 700 to 800 million undernourished by 2010. The two regions projected to have the largest number of undernourished remain South Asia and sub-Saharan Africa. However, while in South Asia their number is forecast to diminish sharply, bringing their share of total population close to the12 percent average of developing countries as a whole, in sub-Saharan Africa the number of undernourished people is projected to increase by more than 100 million to just under 300 million.

The forecast level of undernutrition would exist alongside increasing food imports in the developing countries. Net cereal imports are projected to expand from the nearly 90 million tons of 1989-1991 to some 162 million tons in 2010; and the aggregate cereal self-sufficiency ratio to decline from 92 percent to 90 percent. Although the largest increases are foreseen for the Near East and North Africa (33 million tons) and Latin America and the Caribbean (15 million tons), only a small number of countries in these regions currently face serious foreign exchange shortages.

The doubling in the net cereal trade deficit (from 8.5 to 19 million tons) foreseen for the sub-Saharan region, on the other hand, is more ominous given the precarious balance of payments situation in many of the countries in the region and the unfavourable prospects for many of them, especially those that must continue to finance their growing food import requirements from agricultural export earnings. The above prospects for protracted incidence of undernutrition for hundreds of millions would be the likely consequence of a 'business as usual' approach. To the contrary, all efforts must be mobilised to reduce the incidence of undernutrition and malnutrition as fast and on as broad a geographical scope as possible, so as to achieve by the year 2010 a better outcome than as forecast in the AT2010 study.

The additional amount of food that would be required to increase the per caput consumption of the projected 700-800 million undernourished to the level of average requirements for a healthy life is a minor proportion of world food output. Therefore the issue is not simply whether the world as a whole could produce such an additional amount of food, but even more how to ensure that the countries with the largest concentration of undernourished people improve access to food for all. This would require inter alia a faster increase in the food production rates of those countries projected to have low food supplies and high undernutrition in 2010.

For the developing countries in this class, per caput food supplies are projected to be 2,360 calories in 2010. If none of them were to have less than 2,700 calories by then (which would bring the incidence of undernutrition of the developing countries to a more moderate 6 percent, or 330 million), their production would need to grow at 3.5 rather than the projected 2.7 percent annually. This would barely raise the world growth rate from what is currently projected to 2010 (1.8 percent) to 2.0 percent annually, but what is of great matter is where the additional production is to come from. For those countries and regions with high rates of undernourishment, the task represents a great challenge, as it does in the world community as well. For example, it would imply that in sub-Saharan Africa production growth must be at 4.6 percent for 20 years, instead of 1.9 percent during 1970-1990 and 3 percent projected to the year 2010.

Raising world food production, primarily where natural conditions make it compatible with the sustainability of the natural resource base; ensuring satisfaction at moderate cost of the growing food needs; raising and distributing incomes to enable the largest numbers to provide for their own food needs; providing food assistance to poor and vulnerable population groups; and ensuring stability of food supplies and access, are the objectives which all countries, regions and the international community, have to strongly pursue to prevent dire predictions from materialising.

Food Supply

Annual growth in food production in the region as a whole accelerated from 2.7 percent during the 1970s to 3.0 percent in the 1980s, and further to 3.4 percent during the first half of the 1990s. At the same time, however, annual population growth rates were among the highest in the world—2.7 and 2.9 percent during the 1970s and 1980s, respectively. Consequently, only modestimprovements were recorded in per caput food production for the region as a whole during these two decades. A deceleration in population growth during the first half of the 1990s to 2.7 percent has enhanced the scope for realising gains in per caput food production.

Consumption of food in the region as a whole has been rising rapidly over the last two decades and is projected to continue rising for the foreseeable future. During the 1970s, strong growth in per caput incomes (2.6 percent annually) led to increasing per caput growth in the demand for food (about 2 percent per annum). During the 1980s, however, when the region as a whole experienced negative growth in per caput incomes of some—0.8 percent per annum, per caput food consumption increased by only 0.2 percent per annum.

Changes in lifestyle and dietary patterns over the past three decades have led to a shift in morbidity and mortality rates with diet-related diseases becoming major public health problems, now accounting for more than half of mortality rates in the region. In addition to the rapid rise in the amount of total energy intake, there has been a steep upward trend in the consumption of animal products, fats and sugars and a decline in the consumption of pulses and nuts. Thus, undernutrition of the poor and malnutrition of the affluent exist side by side in some countries of the region.

Overall, in terms of per caput daily dietary energy supplies (DES), which adds together all food items consumed by the population into a common unit, the region had an average of 2 730 calories in 1990-92, among the highest of any developing country region, the average for all developing countries being 2 520 calories during the same period. There were, however, wide variations in DES levels in the different countries. Some of them, including Cyprus, Egypt, Lebanon, Libya, Tunisia, Turkey and the United Arab Emirates, had average DES around 3 300 calories, a level close to the average of the developed countries. By contrast, other countries of the region had average food consumption levels in 1990-92 well below the developing country average, including Afghanistan (1,670), Iraq (2,270), Pakistan (2,340) Somalia (1,590), Sudan (2,150) and Yemen (2,150).

These countries account for more than 75 percent of the number of undernourished people in the region, but important pockets of undernutrition are also found among poor segments of society even in countries with very high average DES levels. Overall, the incidence of chronic undernutrition was estimated by FAO at 16 percent of the total population, compared with 21 percent for the developing countries as a whole. In absolute numbers, this 16 percent corresponds to about 74 million people.

Problems in Food Security

While there are significant differences in resources for food security between individual countries in the region, they share a number of common issues and concerns.

Resource Scarcity

The growing resource scarcities in a relatively harsh physical environment, make exploiting the region's untapped potential and sustaining the quantity and quality of the productive resources, priority concerns. The region is dominated by arid and semi-arid lands with low and erratic rainfall severely limiting food crop production and causing production instability. About 70 percent of the total area is arid or semi-arid, where only expensive irrigated agriculture and limited nomadic grazing are possible. As a result, year-to-year variability in crop production is higher than in all other developing regions.

About 38 million hectares of arable land, approximately 34 percent of the total arable area of the region, is irrigated. The total land area that may be irrigated without excessive investments, about 40 million hectares, is already mostly exploited. Surface water used for irrigation accounts for 90 percent of irrigation while 10 percent comes from ground water sources. Ground water, mostly fossil and thus not replenishable, is of primary importance in several countries.

While further growth in agricultural production from the irrigated areas may not be as high, these areas would require resources to maintain the current level of productivity. Also, further development of irrigation must rely as much as possible on low-cost methods and technologies adapted to local conditions. Examples include water harvesting, use of cheap low-lift pumps and exploitation of shallow aquifers requiring simple methods of water extraction. Also, new approaches to irrigation must emphasise the development of skills and incentives and participatory management of water resources.

Irregular Resource Distribution and Labour Migration

The dichotomy in resource distribution, in particular between oil-exporting and labour surplus countries in the region, represents both a constraint and an opportunity. A major influence affecting economic and agricultural performance in the region has been the labour flow from the capital-deficit to the capital-surplus countries. At its peak in 1985, the number of migrant workers in the oil-exporting countries of the region exceeded 5 million, of which around 3.5 million were from countries within the region.

Although the number of migrants has widely fluctuated, reflecting in particular oil market changes and political developments, remittances from migrant workers continue to represent major sources of foreign exchange. For instance, workers' remittances in 1993 amounted to US$4.96 billion and US$1.56 billion in Egypt and Pakistan, respectively, while the figure for Jordan, US$1.04 billion, is even more impressive relative to the size of the economy. Such flows have contributed significantly to raising rural incomes and, to a certain extent, to rural capital formation.

Returning migrants have, also, brought with them savings and new skills acquired abroad. At the same time, however, the phenomenon has also created labour shortages and reduced agricultural activity in the countries of origin. Labour mobility has also caused added demands on women's already heavy involvement in farm activities. Furthermore, the changing demand for migrant labour in the region has had disruptive effects on current accounts, labour markets and agricultural production in several countries.

Environmental Protection for Food Security

A major issue constraining sustainable agricultural production throughout much of the region is the serious degradation of the natural resources due to soil erosion, desertification, water logging and salinity. In fact, managing natural resources in an efficient and sustainable manner is now one of the most critical issues for food production in the region. Land has been subject to varying degrees of degradation. Wind erosion is believed to affect 35 percent of the total area, while water erosion affects 17 percent. Many of the slopes are stony and denuded by water and wind erosion. This process has led to desertification which is irreversible in many areas.

Where irrigation is intensive, as in Egypt or Iraq, salinisation has emerged as a major problem. In those cases where the major source of land degradation is excessive irrigation, policies that correct for the price distortion of the past, notably free or subsidised irrigation, will help reduce both the problems of excessive water use and its negative effect on soil degradation. It is also important that the agricultural production effort focuses primarily on those areas that offer the greatest potential for cost-effective and ecologically sustainable increases in land productivity.

Market Liberalisation

Almost all countries in the Near East and North Africa have entered a process of market-oriented economic reform, albeit with varying degrees of commitment and success and amidst considerable economic, political and social difficulties in some instances. In certain cases, economicliberalisation has represented a complete break with previous centralised and inward-oriented systems of macroeconomic management. Governments have in general moved away from past practices of large-scale procurement of cereals from producers at high guaranteed prices. Annual increments in support or procurement prices have deliberately been kept low in many countries of the region.

Parastatal monopolies on procurement and marketing have also been lifted and the private sector encouraged to compete with the parastatal in these activities. Similar trends, notably reduced subsidies and increased private sector handling of retailing, are apparent in inputs such as fertilisers. Policy reforms implemented in Egypt in the past four years are a leading example of this process. Key reforms implemented include: ending restrictions on cropping patterns; revoking obligatory deliveries of, inter alia, wheat and rice; freeing up restrictions on private transportation and milling of rice; and rescinding the ban on the marketing and lifting of controls on bread production. A major liberalisation programme along similar lines is in progress in Morocco.

While the reform process is still under way, some countries have already recorded farm productivity and income gains from reduced interventions on farm production and marketing operations, lower trade barriers and freer prices. However, the reduction of previously high levels of both input and food price subsidisation has created the need for special programmes and measures to protect poor farmers, consumers and vulnerable populations.

Role of Susceptible Groups

Even in several countries in the region where chronic

undernutrition on a large scale is not foreseen, appropriate food safety nets are nevertheless needed to protect vulnerable groups in both urban and rural areas. At the same time, there is a need for nutrition education programmes to deal with adverse malnutrition trends of the affluent. The two major food insecure population groups in the region include rural households in both the limited rainfall zones and the highlands, and the urban poor. Safety net programmes in most countries have in the past concentrated largely on the urban poor, based on food price subsidisation policies.

The dominant trend has been one of reductions in general subsidies and price decontrol, implemented within the context of structural adjustment programmes and geared towards reducing high fiscal costs. Cost-effective income transfer schemes are needed so that the limited resources that are made available reach those in need. Targeted food subsidies, food stamps and targeted feeding programmes that identify vulnerable groups and cater for their needs only are to be favoured. Experience has shown that the effectiveness of generalised food subsidies towards ensuring an adequate diet for those in need is questionable (everybody benefits from such subsidies, they are distributionally regressive, they become expensive to maintain and, depending on the level of the subsidy, they may be a distorting factor in the economy).

Commercial Imports of Food Products

The region depends heavily on commercial imports and, in several countries, on food aid to meet its food needs. Net cereal imports increased from 6.5 million tonnes in 1969-1971 to 39.3 million tonnes by 1988-1990 and are projected to reach 75.5 million tons by the year 2010. Along with this considerable increase in imports, the self-sufficiency ratio of cereals for the region declined from 89 percent in 1969-1971 to 71 percent in 1988-1990, and could fall to 68 percent at the horizon year of 2010. Imports of other food items such as livestock products, vegetable oils and sugar have also grown substantially and are expected to continue rising.

As a consequence of such large increases in the volumes of imported foodstuffs, the foreign exchange expenditures on food imports have risen sharply in the past two decades. The value of net food imports into the region increased tenfold from 1969-1971 to 1988-1990 and may double again by the year 2010. Cereals remain the major item of foreign exchange expenditure among imported foods, accounting for 35-39 percent of the total value of food imports. The value of net livestock products recorded an exceptionally large increase of about 800 percent during the period 1970 to 1990 and may register a further 200 percent increase over the period to the year 2010.

The projected large increases in food imports may not constrain the oil-producing countries, but would certainly aggravate the balance of payments difficulties of other countries in the region. Indeed, food imports as a whole amount to some 10-12 percent of merchandise exports. However, low-income countries of the region spend a consistently high share (over 25 percent) of their merchandise export earnings on food imports.

Development of Food Security through Expanded Food Production

For many countries in the region, future economic and food security prospects will be largely contingent upon agricultural performance. Adequate agricultural production growth will be needed to limit food import dependence, meet the growing food needs of urban populations and improve the incomes and food security of poor rural populations, in particular.

Despite resource constraints and often adverse climatic conditions, untapped potential does exist. FAO projects that food output could increase by the year 2010 by over 70 percent as compared with 1988-1990 production levels. Cereal output would grow at an average annual rate of 2.6 percent during the 1990s (wheat by 2.5 percent, rice by 1.8 percent), sugar by 2.9 percent and vegetable oils by 3.6 percent. Growth in output could extend into the first decade of the twenty-first century but at lower rates: 2.2 percent for cereals as a whole, 2.7 percent for sugar and 3.0 percent for vegetable oils.

Growth in food production up to 2010 could be achieved by a combination of increases in yield, intensification of farming and some expansion of the area under cultivation. About 70 percent of the projected increase in crop production over the period is expected to come from yield improvements. This picture runs across the sub-regions with no significant variations between them, except for the Arabian Peninsula, where Saudi Arabia has already reached average yields of more than 5 tons/ha.

This considerable scope for yield improvement reflects the existing wide inter-country yield differential, with most countries at present having average yields well below those achieved by the highest-yield producers. Although many factors are involved in this, the high yields achieved by the better performers suggest that even with the existing technology of high-yielding varieties, improved land and water management, judicious use of modern inputs, especially fertilisers, crop protection inputs, good cultural practices and further adaptive research, there is still considerable scope for increasing yields in a sustainable manner. The adoption of improved technology and farming practices together with efforts to control or eradicate plant pests such as desert locusts, can also be expected to reduce instability in production.

There is also scope for changes in national cropping patterns by introducing new crops such as soya bean and sunflower. Countries like Egypt, the Gulf States, Jordan, Libya and Morocco have made remarkable advances in controlled-environment farming of horticultural products for domestic consumption and the export market. Besides generating foreign exchange and seasonal employment opportunities, such farming systems are noted for their high productivity, controlled use of agricultural chemicals and water saving.

Increases in livestock numbers and in off-take rates would continue to be the dominant sources of growth in meat production in the region. The former is expected to contribute 35 percent and the latter 27 percent of growth in meat production during the coming two decades. Raising production in this manner would require enlarging the feed resource base by increasing the intensity of range and pasture utilisation and by using more feed concentrates and agricultural byproducts. Striking a sustainable balance between livestock numbers and the availability of forage and feed is of paramount importance in the semi-arid conditions of the region.

In some countries, such as Egypt and countries of the West Asia sub-region the lack of potential for expanding the grazing areas means higher yield per animal is an increasingly important source of growth. The intensive and semi-intensive production systems, such as dairy and poultry, which are more responsive to market conditions, are spreading in many countries. Improved production systems and animal health and expanding and upgrading the region's feed base, will all be essential to achieve the projected growth in livestock production.

World Food Summit for Food Security

The World Food Summit (WFS) draft Policy Statement calls on all parties to reaffirm their commitment to policies that will ensure the availability and stability of adequate food supplies as well as access to an adequate diet for all. The WFS draft Global Action Plan stresses the need for each country to choose its own strategy for attaining food security since individual circumstances vary so widely.

The need for differentiated policy responses is particularly evident in the Near East and North Africa region, characterised as it is by marked differences in food security situations and the relative importance of food supply and access in determining them. For a number of low-income countries overall food supply is clearly insufficient to ensure adequate and regular food flows for all and, indeed, a large proportion of their populations are chronically undernourished. For others, aggregate food supplies largely exceed what would be theoretically needed to ensure adequate food intakes for all; yet, significant segments of the population suffer from chronic undernourishment, pointing to the predominance of equity and access problems.

The additional number of calories that would be required to bring every undernourished person in the region to his/her respective energy requirement level is currently estimated at about 586 calories per person/day. This implies that the region's food consumption would have to be higher by 3.2 percent. The cereal component of this higher consumption amounts to approximately 2.6 million tons/year.

It is expected that by the year 2010, the percentage of people chronically undernourished in the region will have fallen to 11 percent, with most countries well below this percentage. Nevertheless, five countries are still forecast to have 20 to 50 percent of their populations chronically undernourished (Afghanistan, Mauritania, Somalia, Sudan and Yemen), while pockets of undernourishment are expected to subsist in the other countries. Appropriate policies must improve this outcome. For the richer countries the objective of altogether eradicating food insecurity is certainly within reach, provided economic and sectoral policies are pursued emphasizing distributional and access aspects.

For the Low-Income Food-Deficit Countries (LIFDCs) in the region, and others where the food security problem is primarily related to inadequate and unstable supply, the obvious necessity will be to enhance domestic food production capacity, focusing on those areas that offer the best opportunities for doing so in an economic and sustainable manner. By combining these policies with appropriate support to the poor and vulnerable groups, even those countries with the highest incidence of undernourishment can expect to reduce it to 10-30 percent by the year 2010.

This improved outcome would also require, however, an abatement of the conflicts and civil unrest which affect several of the countries more severely affected by undernourishment in the region. In all cases, the achievement of food security objectives will depend on the governments' recognition of the fundamental role of agricultural and rural development in addressing the related issues of domestic food supply development, income and employment creation, poverty and food import dependence.

In order to raise each of the regions undernourished persons in 2010 to their respective energy requirement level, an additional 570 calories/day per person would be required. This represents for the whole region, assuming a 60 percent cereal content of this total, 3 million tons/year. This estimate, based on a non realistic targeting of only those undernourished, compares with annual food aid shipments of the order of 1.8 million tons of cereals during the past decade in only one country of the region.

The overall goal for the region would be to reduce the level of undernutrition by a significant margin. FAO considers feasible a reduction in the rates of the undernourished from 16 percent of the total population in 1990-92 to 11 percent by the year 2010. This level of achievement would still leave several countries with high proportions of their populations undernourished (40 or 50 percent in countries like Afghanistan, Somalia and Sudan), and special efforts would be needed in these countries in order to achieve sharper reductions in the numbers of their undernourished people. This would require a further increase in the annual growth rate of food production in the region as a whole (3.1 percent at a minimum, instead of 2.9 percent), but a strong and difficult acceleration in the countries with the highest levels of undernutrition.

It is estimated that the total gross investment for primary agricultural production would have to reach about US$18 billion (1993 United States dollars) annually, of which some 30 percent would be concentrated in the countries with otherwise average food availability below 2 700 calories by the year 2010. To this US$5.0 billion gross of post-production investments should be added, and a further US$6.2 billion would be needed for supporting rural infrastructure and services, which is roughly equivalent to present levels.

Within the framework of the draft Global Plan of Action of the World Food Summit, the priority, regionally-specific actions to achieve such a significant improvement of food security in the Near East and North Africa are as follows:

Enhancing Domestic Food Supply Capacity

Basis for action

There is considerable scope for improving overall productivity, reducing annual and inter-country production variabilities as well as for diversification in crop production. In particular, 14 countries in the region are LIFDCs, for which increasing domestic food production is essential for enhanced national or regional food security.

Objectives

i) expand and diversify food production, especially in the areas with high potential, at least to achieve a 68 percent production increase for the main food crops by 2010 for the region as a whole, and at least by 75 percent for the countries with higher rates of undernutrition (Afghanistan, Mauritania, Pakistan, Somalia, Sudan and Yemen);

ii) strengthen the sustainable management and use of the natural resources, so as to be able to expand areas for crop production by a further seven million hectares (net increase of 7 percent) by 2010; and

iv) enhance the effective, self-reliant participation of farmers and producers in the agricultural and rural development process.

Actions to be taken at the national level

i) secure adequate infrastructural and service support to increase the productivity and output of food crops, especially in the short term, in the high potential areas with irrigation and adequate rainfall;

ii) in all the LIFDCs in the region, but with priority for those with food availability below 2 700 calories per caput/day, launch by the year 2000 programmes targeted at the substantial reduction of the undernourished, including expansion of cereal production by at least 75 percent by 2010;

iii) intensify and diversify food and agricultural production, including cash and fodder crops, livestock and fish culture; the yields of the main cereal crops should increase by at least 44 percent, to over 2.3 tons/ha by 2010. In the countries with food availabilities below 2 700 calories/day, cereal yields should increase to at least 1.8 tons /ha, up from current levels of about 1.3 tons/ha;

iv) expand the reliable availability, to farmers and producers, of an improved quality of production inputs, including irrigation (with an expansion of the irrigated area of at least 10 percent overall, and of 90 percent in countries below 2 700 calories daily intake, by 2010), seeds, plant nutrients (at least doubling chemical fertiliser use by 2010), and post-harvest processing and storage;

v) enhance support services for extension and rural communication, credit and marketing, as well as prevention and control of plant pests and animal diseases;

vi) enhance land tenure arrangements and participatory organisations for farmers, not only for higher productivity but also for more self-reliant management of natural resources for sustainable use;

vii) strengthen national capacity in agricultural research, extension and technology dissemination, especially to support farmers in sustainable technologies for intensification and diversification of production;

viii) promote self-reliant participation by all segments of farmers and producers, both at the national and local levels, with particular attention to poor farmers, and especially women producers;

ix) expand fish production, especially to meet the expected 70 percent increase in demand by 2010 among countries in the North Africa sub-region, by promoting aquaculture and by the better utilisation of small pelagic fish for food;

x) protect, conserve and manage natural resources (soil, water, plant and animal resources), especially through integrated pest management and plant nutrition, by integrating forestry into land management, managing and protecting aquatic environments and fisheries resources. Similarly, monitor and assess the environmental effects of agricultural activities; and

xi) strengthen national capacity for developing and executing policies, plans, programmes and projects with a view to realising sustainable and participatory agricultural and rural development.

At sub-regional and regional levels

i) expand and intensify cooperation, in collaboration

with the International Centre for Agricultural Research in the Dry Areas (ICARDA), in exchanging appropriate technologies and approaches to food production and agricultural rural development, especially in irrigation, crop production technologies and integration of crop production and livestock;

ii) collaborate effectively in establishing and operating joint programmes addressed to transboundary plant pests and animal diseases, such as desert locusts and rinderpest; and

iii) cooperate in the joint management of shared resources, especially by proactive participation in the International Code of Conduct on Responsible Fisheries.

Managing Water Resources

Basis for action

The region's land resources are predominantly characterised by arid or semi-arid conditions, which makes water a critical factor. Low and erratic rainfall severely limits food crop production and causes wide year-to-year variations in agricultural output. The presence of potentially competing claims to limited common water resources makes a regional approach to water development an imperative for many countries of the region.

Objectives

To ensure sustainable and cost-effective use of water resources between various sectors, including an at least 20 percent improvement in the efficient application of water for irrigation and other agricultural purposes by

2010; and to facilitate harmonious management of the shared water resources for sustainable use by the riparian countries and peoples.

Actions to be taken at national level

i) review and reform national water policy and formulate comprehensive strategies for the sustainable use and management of the water resources for various purposes, including the monitoring of water quality and cost-efficiency;

ii) increase the efficiency of water use in agriculture, especially to improve irrigation efficiency through appropriate irrigation technologies, effective water management by farmers, proper usufruct rights and appropriate systems of fees and charges reflecting the true cost of water;

iii) selectively expand irrigation by at least 90 percent by the year 2010 in the countries having a high incidence of undernutrition, and develop low-cost irrigation methods and production techniques with the active participation of farmers;

iv) monitor and survey waterlogged and salinised lands, and reclaim at least 50 percent of these lands;

v) explore the feasibility of, and implement, re-use and ensure the safe disposal of 60-70 percent of municipal and drainage effluent for agricultural production, ground water recharge and desertification control; and promote forestry as part of the strategy for water conservation and land use.

at regional and sub-regional levels

i) expand and strengthen inter-country cooperation for the exchange of technologies and methods for more efficient use of water for food and agricultural production;

ii) promote inter-country negotiation and cooperation in the management of water resources to prevent conflict and uncontrolled ecological damage; and

iii) review and enhance the effectiveness of international river basin agreements and mechanisms.

Meeting Food Import Requirements

Basis for action

The region depends heavily on commercial imports, and several countries on food aid, to meet rapidly expanding food needs.

Objectives

To enhance the capacity of financing food import needs; to balance food import dependence on the one hand, and domestic conditions, cost-efficiency and environmental considerations on the other, including the need to minimise problems linked to supply and price instability through market stabilisation mechanisms and market information and early warning systems.

Actions to be taken at the national level

i) in order to enhance food import capacity, pursue diversified economic growth;

ii) liberalise trade in line with the provisions of the Uruguay Round agreement.

At sub-regional and regional levels

i) develop cost-efficient stabilisation mechanisms at national and regional levels, including national food stocks; promote early warning and market information systems;

ii) pursue, at the international level, the implementation of the Uruguay Round Decision on Measures Concerning the Possible Negative Effects of the Reform Programme on Least-Developed and Net Food-Importing Countries; and

iii) ensure adequate supplies of food aid to food-deficit countries that would otherwise encounter major difficulties in financing the food imports they need.

Pursuing and Deepening Market Liberalisation and Private Sector Involvement

Basis for action

Although market-oriented reforms are being pursued in many countries in the region, the process of market liberalisation, overall and in agriculture, has been uneven. The reduction or elimination of previously high levels of input and food price subsidisation has encountered considerable political and social obstacles, as significant segments of population suffered from the immediate consequences of such measures.

Objectives

i) provide an economic environment conducive to sustained and equitable growth, by stabilising the economies and eliminating costly and distortive interventions and regulations affecting production, marketing and trade; and

ii) ensure an adequate presence of the State in the supply of public goods, infrastructures and social services and in the regulation of competition.

Actions to be taken

i) pursue and intensify market-oriented reforms that address stabilisation and growth objectives while also incorporating social and environmental concerns;

ii) create the conditions for a better supply response to price incentives, through improvements in agricultural infrastructure, training and education and technical support and services to agricultural production;

iii) create an environment conducive to private enterprise development and competitive growth; and

iv) promote private sector investment and broad-based participation in overall and agricultural development.

Enhancing Agricultural Production and Productivity in Highlands and Arid and Semi-arid Areas

Basis for action

Some 70 percent of the areas of the region are arid or semi-arid where low and erratic rainfall severely restrict food crop production and cause production instability. The inhabitants in these areas represent a significant part of the rural poor and are particularly vulnerable to food insecurity.

At the same time, the relatively adverse and fragile agro-ecological conditions demand that food and agricultural development in these areas must be built on sustainable management of the natural resources. These are the challenges confronting most countries in the region, and especially the 14 LIFDCs.

Objectives

To improve the food security of the populations in highlands and arid and semi-arid areas through enhanced, more stable, diversified and sustainable crop and livestock production as well as through development of the off-farm economy.

Actions to be taken at national level

i) develop and introduce farming systems and technologies adapted to more stable, sustainable production in highlands and low rainfall areas, including the use of more drought and disease resistant varieties of food crops;

ii) promote and introduce more sustainable and effective management systems for extensive grazing land, including production of higher-quality fodders within the traditional farming systems;

iii) integrate practical management practices for the protection and conservation of land and water resources, including integration of forestry and trees against desertification;

iv) improve the system for early warning on food shortages and strengthen measures to prevent and control pests and diseases;

v) strengthen agricultural research to provide appropriate technologies for the production systems in these areas, especially to enhance production stability under low-input systems and to enhance sustainable management of land and water;

vi) expand and intensify services for extension and the provision of improved seeds and other inputs;

vii) promote and implement agricultural and rural development schemes targeted at increasing on-farm and off-farm employment, infrastructure for improved services and marketing, as well as at addressing property rights, land use arrangements and participation of poor farmers;

viii) monitor and review the environmental effects of agricultural and rural development, and incorporate lessons in developing strategies and programmes in these areas.

At sub-regional and regional levels

i) exchange and transfer improved technologies and lessons on common problems in enhancing food and agricultural production in the highlands and low rainfall areas; and

ii) cooperate in preventing and controlling outbreaks of transboundary pests and diseases, as well as early warning on food shortages.

Addressing Population and Urbanisation Issues

Basis for action

Population growth rates in the region are among the highest in the world, implying that only modest gains in per caput food production have materialised over the last two decades. This has been accompanied by rapid rates of urbanisation. Even in those countries of the region where aggregate dietary energy supplies are currently adequate, there are sizeable population groups facing food insecurity problems. Considerable scope exists for enhancing broad social participation, particularly for women, in activities relating to food security.

Objectives

i) maintain population and urbanisation growth at rates compatible with the objectives of food security for all;

ii) ensure adequate living conditions and food supplies in urban areas;

iii) ensure access by all groups—including the poorest and most vulnerable—to food necessary for a healthy life; and

iv) create an environment that will enable people to achieve social development in conformity with the principles adopted by the World Summit for Social Development.

Actions to be taken at the national level

i) actively pursue the strategies defined at the Cairo Population and Beijing Conference on Women, as applicable to the country;

ii) provide urban infrastructures and services to ensure that adequate, stable and reasonably priced flows of food reach urban areas;

iii) devise and implement policies to foster employment and access to productive resources by the urban poor, including food-for-work programmes;

iv) provide nutritional assistance and education to the poor segments of the population and those at nutritional risk; and

v) ensure an adequate involvement and empowerment of women in decisional and operational activities related to food supply and distribution.

Promoting Regional Integration

Basis for action

The region is rich in human capital, physical capital and agricultural resources, but these assets are unevenly distributed. Such disparity of resource endowment offers wide areas of resource complementarity and commonality of interests among countries in the region. Nevertheless, progress in implementing regional integration and cooperation schemes has been generally slow and uneven.

Objectives

To exploit the potential for enhanced welfare and food security arising from the complementarity of resources amongst countries in the region by liberalising and intensifying intra-regional trade, in particular of food products, as well as capital and labour flows.

Actions to be taken

i) pursue regional cooperation and integration efforts aimed at enhanced food security, in particular for intra-regional trade in food products and mobility of factors of production;

ii) ensure that regional financial and technical institutions earmark adequate resources for programmes aimed at the improvement of regional food security;

iii) strengthen regional research activity on food production technologies and systems adapted to the conditions of the region; and

iv) foster exchange in technology, know-how and human skills development within the region and in a transmediterranean context.

Responsibilities for implementing priority actions

The national governments have the primary responsibility for creating the conditions required for food security in their countries, through the implementation of the priority actions outlined above. In addition, they have shared responsibilities with other countries within and outside the region, international and non-governmental organisations and civil society at large, in the pursuance of food security goals.

At the regional and sub-regional level

The governments of the region should:

i) promote cooperation among themselves in the exchange and transfer of appropriate technology and approaches, including the establishment of technical cooperation networks among their institutions and exports;

ii) strengthen the cooperation in identification and dissemination of appropriate methods and techniques for cost-effective use of water for agriculture;

iii) share appropriate technologies and approaches for sustainable food production and rural development in highlands and arid and semi-arid areas;

iv) jointly manage shared water resources in the river basins and honour the international agreements they enter for this purpose;

v) extend the scope of national information and early warning systems into a regional network that provides crop and market outlook information;

vi) reinforce regional cooperation mechanisms for food security; and

vii) regional financial, technical assistance and research institutions have responsibility for focusing their activities in favour of regional food security. In particular, regional banks and funds, together with international lending agencies should provide adequate financial support to food-deficit countries in the region that face serious difficulties in financing the food they need.

At the international level

The international organisations and the donor community should:

i) promote and support the national governments and institutions in the planning, execution and reviewing of programmes and projects for food production and agricultural and rural development through technical cooperation and investment;

ii) support, through technical cooperation and investments, national programmes for water resource management as well as inter-country cooperation in water management at the sub-regional and regional level;

iii) support national governments in designing and implementing policies and programmes to control population growth and urbanisation, to enhance food access of poor and vulnerable groups and to provide them with nutritional assistance and education;

iv) contribute where needed, funds for investment particularly in urban infrastructure necessary for food supplies;

v) assist in regional cooperation and integration mechanisms, in particular by facilitating the exchange of expertise and know-how;

vi) promote and support national and regional efforts for enhancing food security in highlands and semi-arid areas. Institutions such as CGIAR will play an important role in fulfilling this need; and

vii) the members of the World Trade Organisation should ensure, within the institutional and regulatory framework provided by that body, the implementation, as applicable to the region, of the Uruguay Round Decision on Measures Concerning the Possible Negative Effects of the Reform Programme on Least-Developed and Net Food-Importing Countries.

The international lending agencies in particular should:

i) provide adequate financial support to food-deficit countries in the region that face serious difficulties in financing their food needs;

ii) assist national governments to pursue and intensify macroeconomic and sectoral economic reforms that pursue stabilisation and structural adjustment along with social objectives.

FAO will play a leading and catalytic role in:

i) providing technical advice and technical cooperation support for expanded food production. It will mobilise its extensive network of cooperation with other international organisations and financial institutions in support of improved food security of the countries in the region;

ii) providing appropriate technical support in the development of policies, strategies and programmes for the agricultural use of water and the related institutional capacity building as well as in the promotion of appropriate technologies in irrigation and technologies for reclaiming water-logged and salinised lands;

iii) assisting, along with other international organisations, in the pursuance of economic reform objectives by providing technical assistance in areas such as: formulation and implementation of policies, strategies and projects for agricultural and rural development in the context of overall reform programmes; building of institutional capacity for policy formulation and implementation; evaluation of the effects of reform programmes on the agricultural sector and the rural poor; financial evaluation of needs involved in reform-related activities;

iv) providing technical advice and cooperation in enhancing sustainable agricultural production and productivity in highlands as well as arid and semi-arid areas by mobilising programmes covering the broad range of issues involved; and

v) providing and channelling technical assistance in the fields of nutritional assessment as well as in services for marketing and supply of food products.

Expected role of civil society

The private sector is expected to play an increasingly important role in taking over functions no longer assumed by the State, in particular activities relating to investment, agricultural production, marketing and trade as well as the provision of inputs. The new emphasis onbroad-based development implies added responsibilities on the part of previously marginalised segments of society, in particular women, smallholders and the poor.

The private sector, non-governmental organisations and interest groups have important roles to play in:

i) technology transfer, investment and skills development through commercial activities;

ii) promotion of, and compliance with, major international agreements such as the International Plant Protection Convention;

iii) sustainable water resources management; in particular the private sector plays a key role in investing in irrigation development;

iv) providing farmers with inputs, services and marketing;

v) undertaking private investment and commercial activities for ensuring adequate and stable food supplies to urban areas;

vi) promoting and supporting programmes catering for the development of arid and semi-arid areas. In particular, non-governmental organisations have many advantages in working with the local populations for their participatory development for food security.

Society at large, including community institutions and local government and non-governmental organisations, has a responsibility for ensuring the involvement and empowerment of women in decisional and operational activities of importance for food security.

4

Soil Biodiversity

Living Organisms in Soils

Soils contain enormous numbers of diverse living organisms assembled in complex and varied communities. Soil biodiversity reflects the variability among living organisms in the soil—ranging from the myriad of invisible microbes, bacteria and fungi to the more familiar macro-fauna such as earthworms and termites. Plant roots can also be considered as soil organisms in view of their symbiotic relationships and interactions with other soil components. These diverse organisms interact with one another and with the various plants and animals in the ecosystem, forming a complex web of biological activity.

Environmental factors, such as temperature, moisture and acidity, as well as anthropogenic actions, in particular, agricultural and forestry management practices, affect to different extents soil biological communities and their functions. Soil organisms contribute a wide range of essential services to the sustainable functioning of all ecosystems. They act as the primary driving agents of: nutrient cycling, regulating the dynamics of soil organic matter, soil carbon sequestration and greenhouse gas emissions; modifying soil physical structure and water regimes; enhancing the amount and efficiency of nutrient acquisition by the vegetation; and enhancing plant health. These services are not only critical to the functioning of natural ecosystems but constitute an important resource for sustainable agricultural systems.

Soils for Agricultural Production

Capturing the benefits of soil biological activity for agricultural production requires adhering to the following ecological principles:

— *Supply of organic matter*. Each type of soil organism occupies a different niche in the web of life and favours a different substrate and nutrient source. Most soil organisms rely on organic matter for food; thus a rich supply and varied source of organic matter will generally support a wider variety of organisms.

— *Increase of plant varieties*. Crops should be mixed and their spatial-temporal distribution varied, to create a greater diversity of niches and resources that stimulate soil biodiversity. For example, diverse habitats support complex mixes of soil organisms, and through crop rotation or intercropping, it is possible to encourage the presence of a wider variety of organisms, improve nutrient cycling and natural processes of pest and disease control.

— *Protection of soil organisms' habitats*. The activity of soil biodiversity can be stimulated by improving soil living conditions, such as aeration, temperature, moisture, and nutrient quantity and quality. In this regard, reduced soil tillage and minimised compaction—and refraining from chemical use—are of particular note.

Improvement in agricultural sustainability requires, alongside effective water and crop management, the optimal use and management of soil fertility and soil physical properties. Both rely on soil biological processes and soil biodiversity. This calls for the widespread adoption of management practices that enhance soil biological activity and thereby build up long-term soil productivity and health.

Adaptation and further development of soil biodiversity management into sustainable land management practices requires solutions that pay adequate consideration to the synergies between the soil ecosystem and its productive capacity and agro-ecosystem health. One practical example of holistic agricultural management systems that promote and enhance agro-ecosystem health, including biodiversity, biological cycles and soil biological activity is organic agriculture.

Organic Management on Soil Biological Activity

Building of soil fertility is the cornerstone of organic agriculture. Organic practices create suitable conditions for soil biotic and abiotic resources through: manipulation of crop rotations and strip-cropping; green manuring and organic fertilisation (animal manure, compost, crop residues); minimum tillage; and avoidance of pesticides and herbicides use. Scientific research in Europe has demonstrated that organically-managed soils significantly increase biological activity and total density and diversity of soil micro-organisms. Such biodiversity enhances nutrients recycling and soil structure. The impact of organic management on soil biological activity and related benefits is summarised below:

— *Abundant earthworms and arthropods.* Organic management increases the abundance and species richness of earthworms and beneficial arthropods living above ground, and thus improves the growth conditions of crops. The biomass of earthworms in organic systems is 30-40 percent higher than in conventional systems, their density even 50-80 percent higher. Compared to the mineral fertiliser system, this difference is even more pronounced. More abundant predators help to control harmful organisms (i.e. pests).

In organic systems, the density and abundance of arthropods, as compared to conventional systems, has up to 100 percent more carabids, 60-70 percent more staphylinids and 70-120 percent more spiders. This difference is explained by prey deficiency due to pesticide influence as well as by a richer weed flora in the standing crop that is less dense than in conventional plots. In the presence of field margins and hedges, beneficial arthropods are further enhanced, as these habitats are essential for overwintering and hibernation.

— *High occurrence of symbionts.* Organic crops profit from root symbioses and are better able to exploit the soil. On average, mycorrhisal colonisation of roots is highest in crops of unfertilised systems, followed by organic systems. Conventional crops have colonisation levels that are 30 percent lower. The most intense mycorrhisal root colonisation is found in grass-clover, followed by the vetch rye intercrop.

Roots of winter wheat are scarcely colonised. Even when all soils are inoculated with active micorrhisae, colonisation is enhanced in organic soil. This indicates that, even at an inoculum in surplus, soil nutrients at elevated levels and plant protection suppress symbiosis. The following are the importance of appropriate living conditions for specific organisms.

— *High occurrence of micro-organisms.* Earthworms work hand-in-hand with fungi, bacteria, and numerous other micro-organisms in soil. In organically managed soils, the activity of these organisms is higher. Micro-organisms in organic soils not only mineralise more actively, but also contribute to the build up of stable soil organic matter. The amount of microbial biomass and decomposition is connected: at high microbial biomass levels, little light fraction material remains undercomposed and vice versa. Thus, nutrients are recycled faster and soil structure is improved.

— *Microbial carbon.* The total mass of micro-organisms in organic systems is 20-40 percent higher than in the conventional system with manure and 60-85 percent higher than in the conventional system without manure. The ratio of microbial carbon to total soil organic carbon is higher in organic systems as compared to conventional systems. The difference is significant at 60 cm depth (at 80 cm depth, no difference is observed). Organic management promotes microbial carbon and thus, soil carbon sequestration potential.

— *Enzymes.* Microbes have activities with important functions in the soil system: soil enzymes indicate these functions. The total activity of micro-organisms can be estimated by measuring the activity of a living cell-associated enzyme such as dehydrogenase. This enzyme plays a major role in the respiratory pathway. Proteases in soil, where most organic N is protein, cleave protein compounds. Phosphatases cleave organic phosphorus compounds and thus provide a link between the plant and the stock of organic phosphorus in the soil. Enzyme activity in organic soils is markedly higher than in conventional soils. Microbial biomass and enzyme activities are closely related to soil acidity and soil organic matter content.

— *Wild flora.* Large organic fields (over 15 ha) feature flora as six times more abundant than conventional fields, including endangered varieties. In organic grassland, the average number of herb species was found to be 25 percent more than in conventional grassland, including some species in decline. Vegetation structure and plant communities in organic grassland are more even and more typical for a specific site than in conventionally managed systems. In particular, field margin strips of organic farms and semi-natural habitats conserve weed species listed as endangered or at risk of extinction. Animal grazing behaviour or routing activity (e.g. pigs) was found important in enhancing plant species composition. Weeds (often sown in strips in organic orchards to reduce the incidence of aphids) influence the diversity and abundance of arthropods and flowering weeds are particularly beneficial to pollinators and parasitoids.

— *High-energy efficiency.* Organic agriculture follows the ecosystem theory of a closed (or semi-closed) nutrient cycle on the farm. Organic land management allows the development of a relatively rich weed-flora as compared to conventional systems. Some "accompanying plants" of a crop are considered useful in organic management. The presence of versatile flora attracts beneficial herbivores and other organisms that improve the nourishment of predatory arthropods.

When comparing diversity and the demand of energy for microbial maintenance, it becomes evident that diverse populations need less energy per unit biomass. A diverse microbial population, as present in the organic field plots, diverts a greater part of the available carbon to microbial growth. This increases the turnover of organic matter, with a faster mineralisation and delivery of plant nutrients. Finally, more organic matter is diverted to build up stable soil humus.

— *Erosion control*: Organic soil management improves soil structure by increasing soil activity and thus, reduces erosion risk. Organic matter has a positive effect on the development and stability of soil structure. Silty and loamy soils profit from organic matter by an enhanced aggregate structure. Organic matter is adsorbed to the charged surfaces of clay minerals. The negative charge decreases with increasing particle size. Silt is very susceptible to erosion since it is not charged, but organic-matter layers on the silt surface also favour aggregates with silt.

Genetic Resources for Food and Agriculture

There are several hundred millions of small farmers in the world who do not have the economic means to buy high yielding seeds or the synthetic fertilisers and pesticides necessary for conventional cultivation. Many of these have opted for the maintenance or re-introduction of organic systems based on traditional forms of agriculture. These promote the use of varieties and breeds that are better adapted to local stress conditions and do not require unavailable inputs such as, for example, veterinary drugs.

There are also farmers who have opted for organic agriculture, in part because they wish to produce healthy and environmentally-friendly food, and also because they are attracted by the strong demand for organic products and the related premium prices. Market driven farmers should, as a minimum, rotate crops as the first step towards improving agricultural biodiversity. This is one of the methods required by organic certification bodies as well as by financial programmes. These farmers have also opted for sowing locally-adapted species and varieties that are more resistant to disease and local environmental conditions because synthetic fertilisers and pesticides cannot be relied upon.

The adoption of organic agriculture methods requires farmers to follow a series of agronomic practices (e.g. crop rotations, crop associations, green manure and maintenance of vegetation between rows) that make organically managed systems biologically much more complex than conventionally managed systems. Organic farms make use of larger numbers of plant and animal species than conventional systems. As a result, the large pool of genetic resources for food is maintained and other useful organisms, such as predators, pollinators and soil micro-organisms are increased—for the very benefit of the agricultural system.

The contribution of organic agriculture to the survival of centres of diversity and to under-utilised species, varieties and breeds is described below. The specific requirements of organic farmers for a productive genetic material, in conditions of low-input and harsh environments, are generating innovative approaches to research and development; the emergence of participatory systems of selection and distribution of appropriate genetic material is presented.

Maintenance Genetic Diversity

The continued cultivation of crop species within their centres of diversity plays a fundamental role in the maintenance of genetic diversity. Preserving the integrity of centres of diversity through ecologically-sound agriculture is an indispensable inheritance for agriculture and as such, for food security for future generations. It is the genetic variability that allows populations to adapt to changing environmental conditions.

In centres of diversity, the introduction of organic practices is aiding the conservation, through cultivation, of populations with high genetic variability. The maintenance of agricultural production in centres of diversity requires market outlets for peasants and indigenous communities. The new income opportunities offered by organic markets reverse the present trend of abandoning land which has previously been economically uncompetitive.

Examples where organic agriculture created viable means for in situ conservation and use of areas with diverse genetic heritage include: producing and processing cocoa in Mexico to sell Maya chocolate to tourists, and the maintenance of naturally-pigmented cotton in Peru, resistant to pests, diseases and drought. These organic market opportunities provide for the economic viability of thousands of farmers and processors and maintain genetic variability for future generations.

In the past, agriculture has played an important role in the maintenance of genetic diversity. The substitution of a large quantity of species for only a few and the adoption of high yielding and uniform varieties from a genetic point of view, has caused a significant reduction in the genetic inheritance of cultivated species. Many agricultural species, varieties and breeds which have played an important role in the human diet and traditional cultures have practically disappeared over the last century.

Organic farmers breed varieties for quality, nutrition, resistance and yield, in reduced input growing conditions. Research has shown that these characteristics are more likely to be found in older native cultivars. In particular, open pollinated varieties and indigenous breeds offer diverse and regionally adapted characteristics that are better suited to organic agriculture. In the last decade, the adoption of organic agriculture has indirectly established a rescue process of species, varieties and breeds threatened by under-use or extinction. The restoration and enhancement of under-utilised species and varieties has been motivated by specific demands of both consumers and farmers.

Development of Market

For the rescue of varieties threatened by extinction, the development of a market is fundamental and it is here that organic agriculture plays an important role as the price premium gives an additional value to the product. This is especially the case now that there is a consumer' interest in traditional, speciality and organic products. Many consumers search for quality aspects, for either health reasons (e.g. gluten-free crops, other medicinal properties or high fibre content) or culinary traditions (e.g. gastronomy, taste and local diets).

Examples where organic agriculture has allowed the maintenance and improvement of species and varieties that otherwise would suffer strong genetic erosion or extinction include: the discovery of the nutritional value of the gluten-free quinoa in Peru and saraceno grain in Italy; re-introduction of local rice varieties in traditional diets and culture in Indonesia; and economic viability of the Garfagnana spelt in Italy. These cases provided for the survival of poor communities in marginal areas and valorised endangered genetic resources.

Productivity under low-input conditions

The majority of crop varieties available on the commercial market are not suitable for organic cultivation methods as they have been selected for production dependent on irrigation and large quantities of synthetic fertilisers and pesticides. Many of these are hybrids and are not open-pollinated. In the last few years, the problem has worsened following the arrival on the market of genetically modified varieties.

The selection objectives of organic agriculture differ from those for conventional agriculture. It is of crucial importance to utilise the genotype potential for an improved adaptation of varieties to the low-input conditions prevailing in organic agriculture. The necessity for organic farmers to find methods for obtaining quality products with good yields and limited production costs is greater than for other farmers. Besides the fact that organic farmers cannot apply synthetic inputs, their use of organic fertilisers, natural pesticides and other permitted substances is uneconomical in the long-term.

Permitted external inputs as such are relied upon mainly during the conversion period to organic agriculture or under exceptional circumstances. The comparative advantage of certain varieties to withstand local natural stress, especially in marginal areas, leads organic farmers to adopt biodiversity management as their main productive strategy.

Empirical organic breeding systems have been based on the selection of individuals better adapted to the local environment and that are more resistant to pests and diseases. Many of these systems have demonstrated interesting results in restoring and improving local varieties. Examples of restoration of varieties and breeds include: the rescue, in Germany, of an old variety of wheat with a vegetative cycle that allows the absorbency of nitrogen available in sandy soils; in Cuba, the success of local pumpkin varieties is used as the basis for the refinement of methodologies for the selection of varieties for low-input situations.

Environmental conditions

Animal breeding for high performance and selecting for early maturity have led to increased susceptibility to diseases, joint inflammation and mastitis as well as circulatory, metabolic and fertility problems of livestock. Loss of breeds is exacerbated by the narrowing genetic base of modern breeds and hybrid lines. The trend towards inbreeding increases the degree of genetic uniformity in the animals and hence, susceptibility to infection, parasites or epidemics.

A significant proportion of local breeds remains in the care of pastoral people and traditional livestock owners in developing countries (e.g. pigs in China, cows in India and poultry in Asia and Latin America). Local breeds are suitable for free ranging and robust, thus viable in marginal areas. While the yield may be less in the short-term, animals are more resilient and able to survive in the long-run.

Examples where organic agriculture restored, through utilisation, genetic resources resilient to local natural stress include: productive rearing, in Italy, of authochthonous races of the Maremmana cattle, on the verge of extinction, due to its suitability to grow in marshy environments, and the re-establishment of native poultry in South Africa, due to their resistance to Newcastle disease.

Selection and Distribution of Organic Genetic Resources

Historically, farmers have managed many varieties and breeds according to agronomic and culinary properties. Considering the need for a wide gene pool to improve and multiply genetic resources for food and agriculture, breeding requires access to seeds and breeds from the formal and informal sectors. Open pollinated varieties, which represent an important gene pool for resource-poor farmers living in marginalised and stress-prone areas, are rapidly vanishing. They are replaced by very few hybrid varieties which require inputs not available to poor farmers and which entail dependence on large seed companies.

Limitations and threats associated with crops have stimulated many organic farmers, especially in the horticulture sector, to produce their own seeds. In order to do this, they have often had to rescue local varieties and develop their own system of selection and distribution. In many cases, the systems include the exchange of seeds between farmers as a fundamental instrument (e.g. organic seed fairs).

Organic systems encourage the preservation and expansion of older, locally bred and indigenous varieties and breeds. Farmers who save their own seeds can gradually increase crop resistance to pests and diseases by breeding for "horizontal resistance". Horizontal resistance is the ability of a crop to resist many or all strains of a particular pest (which differs from breeding for "vertical resistance" to have a gene to resist one specific strain of a disease).

By exposing a population of plants to a certain disease or pest (or to several pests at one time), then selecting a group of the most resistant plants and interbreeding them for several generations, a given population becomes more resistant than the original population. Horizontally resistant cultivars are well adapted to the environment in which they were bred, but may be less suitable for other growing conditions.

Benefits derived from new varieties bred by farmers require a legal system of common ownership that allows equitable access and benefit sharing. The biodynamic network of farmers and breeders in Germany provides an example of how such a system could be organised: trials, selection and evaluation of genotypes adapted to low-input conditions is made by farmers and common ownership of new varieties is shared by the community.

Organic agriculture is providing an important contribution to the in situ conservation, restoration and maintenance of agricultural biodiversity. The spontaneous establishment of participatory systems of research and development is shaping a simple and practical system of equitable sharing of benefits derived from genetic resources for food and agriculture. The growth pattern shown by the conversion to organic agriculture throughout the world suggests that this contribution is likely to increase still further.

Wildlife and Agricultural Biodiversity

According to the IUCN Red List of 2000, approximately 70 percent of all endangered species of birds and 49 percent of all plant species are spoiled by agricultural activities and approximately 25 percent of the world's wild animals and plants is running the risk of extinction by the middle of this century. Agricultural productivity depends upon the maintenance of ecological balances and the natural properties of plants and animals. The fundamental role of maintaining (or restoring) biodiversity is demonstrated through ecological services such as pollination of crops, predation for biological control of pests, micro-organisms' maintenance of soil fertility and other services vital to the food web.

On the other hand, agriculture has the same important role in wildlife conservation, provided that it avoids the use of substances (e.g. pesticides) that could have a harmful effect on natural species and that it maintains food and shelter through a diversified landscape and permanent vegetation (e.g. trees, hedges and fields margins). Finally, a type of land use that provides suitable biological corridors is essential for wildlife conservation.

Nature conservation has traditionally consisted of geographically targeted efforts. While this approach has resulted in a number of successes for rare species or key locations, worrying declines of protected species have occurred. A healthy environment is a prime objective for the conservation of vital terrestrial ecosystems and the wildlife in it. Natural faunal and floral species have strong connections with agriculture, whatever their habitats are, especially as agricultural fields occupy much of the earth's land surface.

Protected areas simply cannot be viewed in isolation from the communities within and near them. Almost everywhere, rural dwellers claim historical rights on protected areas which governments have, at a point in time, declared "protected" for national interest. People inhabiting within or in the neighbourhood of protected areas depend directly on their resources for a living. In India, for example, at least three million people live in protected areas and many other millions live in their proximity.

In Latin America, about 86 percent of national parks are inhabited by indigenous people and migrants. This very dependency on protected areas and its diversity of life forms imposes ecological farming policies. If nature is to be protected successfully, protected area dwellers should be given agricultural choices which are not environmentally destructive and economically rewarding. Considering that the relationship between wild biodiversity and agriculture is reciprocal, the protection of wildlife, biodiversity and natural areas must include a correct management of agricultural systems.

Agricultural Biodiversity and Natural Conservation

There is no doubt that farmers are the most important managers of natural resources. Several studies indicate that organic agriculture safeguards non-agricultural biodiversity and offers a viable alternative in protected area categories where human activities are allowed. Most importantly, the huge land surface surrounding protected areas requires an agro-ecosystem management that preserves the safety and integrity of the landscape. If farm land bordering and connecting protected areas employ organic methods, there is no reason to fear the loss of wildlife or contamination of air, water and soil.

These buffer zones are critical to the success of conservation in the protected areas.

Organic agriculture enhances people's ability to live in harmony with nature and to derive economic benefit from their land. Considering that most protected areas traditionally belonged to local villagers, organic agriculture allows local people to maintain some control over their land, protect land and biodiversity through their farming practices, reap its benefits for themselves and, at the same time, conserve and improve the natural environment.

The direct impact that organic agriculture has on ecosystems can be seen on different scales: on-farm, farm margins, and overall ecosystem. While on-farm biodiversity has been discussed in the sections above, the following sections will consider the interactions of organic agriculture with the wider landscape, namely protected areas and buffer zones.

Organic agriculture in protected areas

Certain protected area categories allow sustainable land-use activities such as organic agriculture, management of non-timber forest products, fishing, subsistence hunting and ecotourism. Organic farming within protected areas is a growing practice that can "help to define and control sustainable land uses in those protected areas containing significant human populations". There are some 350 Biosphere Reserves in 85 countries to protect ecosystems. The area surrounding a biosphere reserve, named buffer zone, plays a critical role because activities carried out here strongly influence the core of the protected area itself. Often buffer zones are areas dedicated to agricultural practices.

Conversion to organic systems can reduce the detrimental effects of conventional farming, and can provide sustainable systems, suitable for the management restrictions governing buffer zones, and consequently to natural ecosystem conservation. The protection of the natural heritage must consider the impact, be it positive or negative, that human activities have on it. The characteristics of agriculture make it one of the main activities to be practised in protected areas and buffer zones.

Organic agriculture offers a suitable alternative in ecosystems where geographic and morphological conditions are favourable to human activities, such as wetlands and lowland forests. Also, the ecological services offered by organic agriculture in biological corridors are of extreme importance.

Wetlands Conservation

Wetlands are defined in the Ramsar Convention as "areas where water is the primary factor controlling the environment and the associated plant and animal life. They occur where the water table is at or near the surface of the land, or where the land is covered by shallow water". Wetlands are present in every country, from the tundra to the tropics. Their ecological importance derives from their capacity to host high concentrations of birds, mammals, reptiles, amphibians, fish and invertebrate species. In fact, wetlands are boundary areas that combine the components of marine, fluvial and terrestrial ecosystems.

In addition, many functions of wetlands derive from the interactions of different properties of soils, water, plants and animals. Some of these functions include, for example, water purification, water storage, flood mitigation, recharge and discharge of underground aquifers (by the movement of water), and stabilisation of local climatic conditions. Natural wetlands are among the most threatened ecosystems in the world. Their high productivity and the morphologic characteristics make wetlands excellent areas for many human activities, especially agriculture. Often the damage caused by land reclamation for agriculture or unsustainable practices leads to the disappearance of wetland areas and corresponding biodiversity. Suitable management is therefore a critical priority to save these fragile ecosystems.

Organic agriculture can help wetland conservation, by providing suitable habitats for wildlife species, reducing water pollution and, at the same time, offering a valid economic alternative to the exploitation of natural resources. Examples where the avoidance of synthetic inputs and cropping strategies employed by organic farmers in wetlands offered breeding and feeding habitats for endangered wildlife include: return of cranes and storks in the cereal production in Muravia Park, Russia and of egret and heron in rice production in the El Ebro Delta, Spain.

Organic beef production in the Pantanal Region, Brazil, created natural grasslands vital to wild mammal herbivores in an area considered the world's biggest ecological sanctuary (including many endangered species such as the Pantanal marsh deer) and which previously suffered from deforestation of savannah and implementation of artificial pastures for beef.

Human activities such as modern agriculture and grazing can be a serious threat to forest ecosystems and hence, for a large percentage of the world's flora and fauna. In fact, the practice of clearing the tree cover often interrupts the continuity of canopy, an essential characteristic for genetic and specific biodiversity flow. In many cases, the same agricultural fields are also a barrier to wildlife movement, and pollution from agrochemical abuse may have negative impacts on forests.

Organic management in forest areas has the potential to reduce the loss of biodiversity caused by these agricultural activities. The organic systems utilised in forest areas are shade and sun cultivations. In the first case, the plants are grown under a canopy that consists of original forest trees or selected plants. In the latter, there is no forest cover on the agricultural field. Although yields of full sun production are often higher than for shade cultivations, the latter provide timber, fuelwood and other fruits, and have a lower soil erosion rate. Shade plantations have been shown to be highly beneficial to biodiversity conservation in tropical forest ecosystems, including millions of migratory birds and other animals and plants.

For example, the Smithsonian Migratory Bird Centre's research has demonstrated that bird species are almost twice as large in shade coffee as in sun coffee. Shade coffee cultivation therefore offers important conservation opportunities as the structural profile of shade coffee farms is similar to natural forest, providing habitat suitable for resident birds and migrants. As a result many scientists and conservationists believe that shade coffee plantations are ideal zones for migratory birds.

The Northwest Shade Coffee Campaign has obtained statistical data on shade coffee as compared to sun coffee and has shown how shade cultivations are preferred by bird species (approximately 150 compared with 20-50 species), mid-size mammals (24 species compared with almost none) and several species of ants, beetles, amphibians, epiphytes and others. Smithsonian Migratory Bird Centre's studies indicates that at least 180 species of birds live in Mexican shade coffee and cocoa fields (much more than on other agricultural lands), and that 90 percent fewer bird species live in Colombian sun coffee plantations than in shade coffee.

Researchers highlighted the capacity of the canopy cover to support secondary structures (e.g. epiphytes, parasites, mosses and lichens), which in turn support arthropods, amphibians and other living beings. Another relevant fact is that shade coffee hosts a large density and diversity of predators and parasitoids involved in the control of insect pests. Shade coffee offers an optimal habitat also for other tropical forest species, such as beetles, ants, wasps and spiders, and supports a high diversity of many vertebrate groups (small mammals as opossums, squirrels, mice and bats). Finally, areas of high ecological value located around coffee farms are protected.

Examples of the co-existence of shade organic cultivations and richness of biodiversity in forest ecosystems include: shade coffee in the buffer zones of El Trionfo Biosphere Reserve, Mexico; shade cacao in the tropical forest of Montes Azules Biosphere Reserve, Mexico; and yerba mate in the threatened Atlantic rainforest of the Guayaki Biological Reserve, Paraguay. Examples of conversion to organic agriculture to reduce pressure on endangered forests include: land exchange between Del Oro orange production and Guanacaste Conservation Area in Costa Rica to restore rare native forests; organic bananas to protect the threatened Guaraquecaba Atlantic forest and its extraordinary biodiversity in Brazil; and the re-introduction of native species through agroforestry in Ampay Forest Sanctuary in Peru. In all these cases, local communities' income was raised while providing benefits to the environment.

Biological corridors

Corridor zones link protected areas with one another and either remains under wild cover, or are managed to ensure that human land use is compatible with the maintenance of a high degree of biological connectivity. These areas have a fundamental ecosystem function forming part of a large-scale ecological web. To be effective, biological corridors must offer suitable habitats to wildlife, thus the maintenance of a healthy environment is necessary. Areas joining parks or reserves are equally important for biodiversity conservation as they ensure ecosystem connectivity. These "linkage" areas often host agricultural systems but if they are managed in an unsustainable manner, the vitality of these corridors decreases or is totally hindered.

Careful organic management in agricultural land between (and sometimes inside) the protected areas has permitted the creation of important biological corridors and allowed the protection and increase of biological diversity. Through on-farm structured vegetation and canopy tree diversity, organic agriculture has demonstrated its capacity to provide livelihoods to farmers while providing easier movement of animals between managed forests and protected areas. One of the largest, most unique conservation efforts presently underway in the world is the Meso-American Biological Corridor that runs across seven countries, covering an area of 770 000 km^2, from Panama to Mexico. Examples of organic cultivations include: shade coffee in the buffer zone between El Imposible and Los Volcanes National Parks in El Salvador where hundreds of endemic and threatened birds, mammals, reptiles and trees strive; and shade cacao and bananas in the Talamanca-Caribbean Biological Corridor and its buffer zones, Costa Rica, where most of the fauna is endemic.

Organic systems offer habitats similar to intact forests, providing suitable refuge for migratory and resident birds and foraging for many forest mammals, including species at risk. Organic land management and bird conservation programmes exist in several northern countries. Examples include: organic crop and livestock production in Brandenburg, Germany where the higher presence of skylarks has been observed; and organic farming inside a wildlife corridor that connects Adirondak National Park and Lake Champlain in the State of New York, USA.

5

Soil and Water Problems

Soil and water quality problems caused by agricultural production practices are receiving increased national attention and are now perceived by society as environmental problems comparable to other national environmental problems such as air quality and the release of toxic pollutants from industrial sources. Severe soil degradation from erosion, compaction, or salinisation can destroy the productive capacity of the soil and exacerbate water pollution from sediment and agricultural chemicals. Sediments from eroded croplands interfere with the use of waterbodies for transportation; threaten investments made in dams, locks, reservoirs, and other developments; and degrade aquatic ecosystems.

Nutrients accelerate the rate of eutrophication of lakes, streams, and estuaries; and nitrogen in the form of nitrates can cause health problems if ingested by humans in drinking water. Pesticides in drinking water can become a human health concern and have been suggested to disrupt aquatic ecosystems. Salts can be toxic at high enough levels and can seriously reduce the uses to which water can be put. In some areas, toxic trace elements in irrigation drainage water have caused serious damage to fish, wildlife, and aquatic ecosystems.

In US, renewed concern about soil erosion led to major new initiatives in the 1985 Food Security Act. For the first time, to be eligible for farm programme benefits, agricultural producers were required to implement a soil conservation plan for their highly erodible croplands. A conservation plan was required for highly erodible land converted to cropland, and Congress also established the Conservation Reserve Programme to pay producers to take highly erodible land out of production.

Sheet and rill erosion remains an important problem, causing soil degradation on about 25 percent of U.S. croplands (Figure 1).

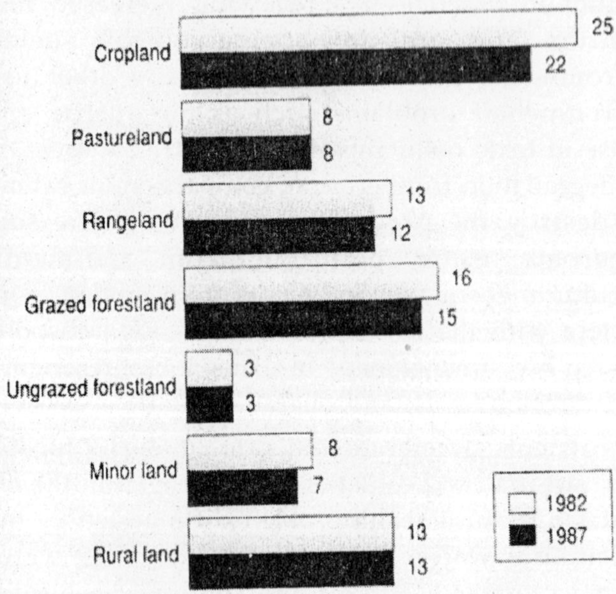

Figure 1. Percentage of land eroding by sheet and rill erosion at greater than the soil loss tolerance level.

Other forms of erosion—such as wind, gully, and ephemeral gully erosion—are also important and, if quantified, would expand the reported area of cropland on which erosion causes soil degradation. Conservation Compliance and Sodbuster, which are provisions of the 1985 Food Security Act, should result in substantial reductions in erosion caused by both wind and water. If these provisions are fully implemented and if the conservation practices remain in place, the United States will have taken a large step toward solving a soil erosion problem that has plagued U.S. agriculture since settlement by Europeans began.

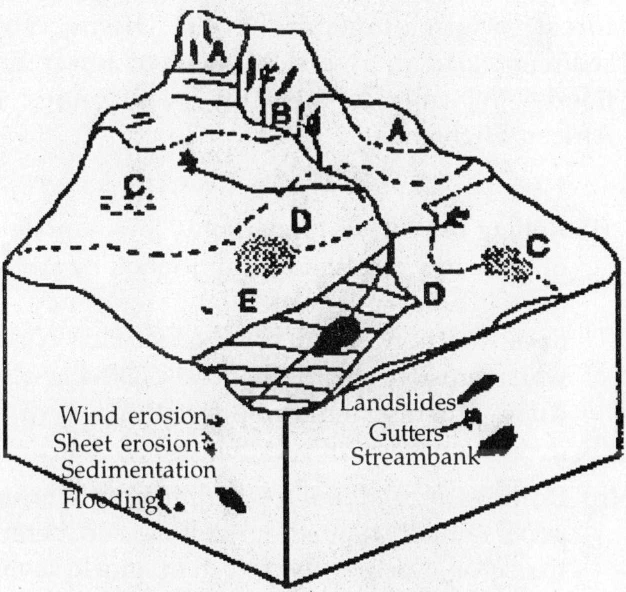

Figure 2. An Idealised Catchment

Even as major strides toward erosion control are being taken, however, new concerns about the soil resource are emerging. Compaction is increasingly noted as a factor that degrades soils and reduces crop yields, but no comprehensive data on the extent or severity of compaction are available. Salinisation of soils, particularly in the western part of the United States, is causing serious and often irreversible damage where it is occurring.

One of the advantages of a watershed approach is that it allows the spatial or geographical linkages to be readily identified. The idealised diagram in Figure 2 suggests that the following broadly defined terrain types can be used:

a) Steep uplands, high rainfall, fast rivers, deeply dissected, naturally forested, initially low population density. Problems: erosion, loss of forest, origin of major floods. Offsite effects: sediment load in rivers, siltation of lower dams, floods. Typically in the Himalayas, Andes, East African Highlands.

b)

(i) Rolling or plateau topography, low rainfall, arid or semiarid vegetation, rangelands or irrigated agriculture. Problems: tendency to desertification, wind erosion, surface crusting, water erosion and flash floods. Off-site effects: dust storms, channel erosion, sediment deposition. Typically Indus valley, Sahel.

(ii) Rolling or plateau topography, moderate to good rainfall, rainfed agriculture and perennial tree crops, originally forest or moist savanna. Problems: acidity, poor structure and nutrient content of soils, poor distribution of rains, erosion under poor management. Typically West Africa,

c)

(i) Flat topography, low rainfall, very small fall on river, naturally arid, irrigated agriculture Problems: siltation of river, unstable channel, floods, salt or sodium contents of soils. Off-site effects: pollution of ground water, saline downstream river. Typically lower Indus valley, Euphrates - Tigris rivers, lower Nile.

(ii) Flat topography, high or moderate rainfall or high watertable, naturally forested. Problems: floods, impoverished and acid soils. Off-site effects: few. Typically lower Niger,

(d) Wetlands, with high watertables, flooded for part of the year. Often used for irrigated agriculture, with high-yielding, deep-water rice, and aquaculture. Problems: rice yields appear to be falling, for unknown reasons, and environmental problems are caused by methane production and potential impacts of sea level rise.

Nutrient Cycle and the Chemical Fertility of the Soil

There are two, perhaps three, basic cycles in which the main plant nutrients circulate. The first consists of the uptake of nutrients from the soil, their incorporation into plant tissue, and their direct return to the soil. This natural cycle dominates in almost all natural vegetation, and is one of the biophysical bases of shifting cultivation. In agriculture there is an additional loop, in that nutrients may be removed in harvested material, but partly returned in human or animal wastes.

In far too many parts of the less-developed world this nutrient cycle is no longer efficient, because the duration of the fallow period has become too short. Finally, all vegetation has an external open "cycle", where nutrients escape to the atmosphere or to rivers or groundwater, and are indirectly and partly replaced from the atmosphere or soil mineral breakdown. This external "cycle" increases greatly during soil degradation. It is also a main component of soil acidification, which is due to loss of cations from the soil profile.

Where the social and economic conditions are such that nutrients in harvested produce are not returned to the field because it is sold to cities or abroad, the harvested nutrients become part of the external cycle also. The removal in the harvest from intensive agriculture can be very large, up to over 200 kg/ha of N or K. It is then necessary to replace these losses with fertiliser. Such heavy cycling is difficult to control accurately, which is the main reason for the losses in this type of agriculture. Considerable advances have been made in modelling movement of N in plants and soils over the past 20 years, so that control can now be improved.

A key requirement is that nutrients, particularly nitrogen, shall be supplied as far as possible at the time when they are required by the growth of the plant, so that they are taken up relatively rapidly. This is particularly important where heavy rainfall and light-textured soils lead rapidly to leaching. However, the unknown local variations in growth, yield and naturally present soil nitrogen mean that precise control is still difficult. The decision on the correct amount of fertiliser for a particular piece of land under a particular crop is still very open to error, even in highly developed agriculture.

The very low fertility of soils under shifting cultivation found in much developing-country agriculture mean that yields and nutrient offtakes are correspondingly low, and are often decreasing (Table 1). As the fallow period shortens, the internal nutrient cycle becomes still less effective. It is imperative to import nutrients as manure, residues or fertilisers in such cases. There are many ways in which the nutrient cycle can be more tightly closed, such as by recycling all plant, animal and human residues, applying hedgerow or agroforestry prunings, and the use of legume crops, but this does not make the soil productive if it starts from a severely depleted state.

Even the best-managed cropping will have some losses through the external cycle, which have to be replaced. Nitrogen is a special case because it can be fixed from the atmosphere, though this is often not sufficient to meet crop needs. Cropping of a soil which is continually losing nutrients, particularly phosphorus and potassium, is ultimately non-sustainable. This is the basic argument for the use of fertilisers. The challenge is to use these for intensification so skilfully that the pollution problems seen in temperate (and some tropical) agriculture are avoided, and that their use becomes sustainable.

It is even more challenging to do so in economical and affordable ways for resource-poor farmers with small farms, especially in Africa. Nutritional problems are often severe in shifting cultivation systems as these become more intensive. The way in which productivity decreases as the system moves from Phase 1 to more intensive states depends greatly upon the soil type and its resilience. Very few soils will stand continuous removal of nutrients in continuous cultivation.

Table 1 contains some excessively low yield values, that are presumably largely due to nutrient depletion. Nutrients contained in organic residues are very valuable, and should be used as extensively as possible, both to enhance nutrient cycling, and because they have slow-release characteristics that may in some circumstances make them moie efficient than fertilisers for the same total amount of nutrient. Their prime drawback is that the quantity is limited.

Table 1. Yield per Unit Area (kg/ha) of Some Major Staples in Some African Countries

Country	Crop	1961/65	1969/71	1979/81	1989/91	1993
Côte d'Ivoire	Rice	890	1168	1171	1174	1334
	Maize	680	773	700	713	831
	Cassava	2500	3300	5266	5680	4521
	Sorghum	500	507	538	563	600
Ghana	Maize	550	982	1078	1300	1509
	Cassava	8500	7419	8647		7226
Kenya	Wheat	1678	1678	2011	1747	1579
	Maize	1100	1489	1360	1300	1249
	R&T		7722	7993		8077
Niger	Millet		422	435	383	357
	Sorghum	480	445	432	280	153
Nigeria	Maize	920	869	1350	1300	
	Sorghum	850	652	634	1093	967
	R&T		9585	9728	9930	10193
	Cassava	6200	10592		11150	10500
Sudan	Millet		567	397	166	
	Sorghum	900	808	725	534	509
Tanzania	Maize	1180	813	1306	1340	1404
	Sorghum	870	503	763	970	1102
	R&T		4902	9491	8280	8223
	Cassava	4100	4854	12071	10830	10400
Uganda	Cereals		1069	1551		1528
	R&T		4449	5802		6594
Zaire	R&T	12100	6795	6901	7562	7906

It is argued above that higher production will eventually require the use of fertilisers, which are sometimes considered suspect on grounds of sustainability. Potassium and phosphorus are produced by extractive industries, and their supply is therefore not indefinitely sustainable. In fact, the potential reserves of potassium are about 1000 times greater than the annual production, and phosphate rock resources are about 450 times greater.

Also, nutrients from fertilisers may escape from the local nutrient cycle, and cause pollution of water ways and groundwater, with eutrophication and health hazards. This problem can be largely controlled by careful agronomy, and this must become a larger research topic in less-developed countries in the future.

This consideration of nutrient cycles and balances suggests that greater efficiency in uptake by plants is of only partial utility, because nutrient must at some stage be resupplied if it is to be taken off in crops. However, more efficient uptake does allow the soil in a field to be operated at a lower nutrient concentration than otherwise, and this will tend to minimise losses. Low soil concentrations of nitrogen also encourage resupply of nitrogen by associative or symbiotic fixation, which can supply all or part of the requirement. Associative fixation had an uncertain record at one time, but it appears that some grasses can fix important amounts. The fungal mycorrhizal root associations can improve the uptake ability for phosphorus, and thereby make more of the soil phosphorus available at a given plant growth rate.

The worst understood part of soil science and plant nutrition today is almost certainly the soil microbial population and its functions in relation to higher plants. Very few of the microbial species are identified, the systematics are fragmentary, and the functions are known accurately only for a limited number of specialist organisms. Many of the soil processes that they carry out are essential for plant nutrition, soil structural stability, and several of the most important biogeochemical cycles. Considerable strides have been made in the past decade, but the introduction of new molecular biology techniques should allow accurate identification of both species and function in the future.

Physical Protection of the Soil

Soil structure is still not well understood at a fundamental level, and much of the science is largely empirical. It is still not possible to describe 3-dimensional structure or porosity in fundamental terms, and a great deal of reliance has to be placed on inferences from the moisture characteristic and the hydraulic conductivity. The strength and impedance of soils are closely allied to structure and texture, but cannot be predicted in a mechanistic way.

The quantity and behaviour of the particular clays in the soils (low-activity kaolinites or active smectites) together with the amount of soil organic matter are critical. In particular, it is difficult to predict the degree of resilience of a soil following structural damage. This rather weak theoretical basis means that much of the work in these subjects must be empirical, applied and adaptive, though absolutely essential.

It is extremely important to preserve structure, otherwise capping, loss of infiltration, wind and water erosion, and root impedance easily occur. The general field methods of preserving structure and porosity are well known from long experience; maintain soil organic matter levels, maintain vigorous plant growth, prevent heavy loads, prevent large raindrop impacts at terminal velocity, and cultivate with care at the right time. Even so, the immediate strength of the soil structure, and the speed with which it regains structure after damage (resilience) are not properly understood.

Even soils with similar textures may show great differences in the persistence of structural damage. Research must therefore aim at better practical and site-specific methods of preventing damage, or remedying it in suitable cases, and a great deal of progress has been made in this way. Social aspects are important. For example, even when the value of mulching is recognised for structure protection, there are often alternative uses for the vegetable material which have priority.

Wind erosion can be devastating, and is a critical component of the loosely defined process called desertification. Once soil movement starts it helps to loosen more soil, and dust storms rapidly develop. The essential control measure is to maintain a sufficiently dense vegetation cover, so that soil does not start to move, which is why intensive grazing is so dangerous, especially under variable rainfall regimes and socioeconomically driven high herd and flock numbers.

Water erosion can occur as sheet, rill, gully or river-bank erosion, depending upon the topography, soil type, rainfall intensity and erosivity, infiltration capacity and length of run. Some erosion losses always occur, even under natural conditions, and less than 2 t/ha/y are usually regarded as acceptable, depending upon the assessed rate of formation of fresh soil at the bottom of the soil profile. However, in serious cases the loss may be over 100 t/ha/y. The provision of planted or natural strips of permanent vegetation along field contours and river banks is useful in lessening river silt load, though the latter may not affect the actual erosive process on the field itself.

A variety of erosion models exist, but the whole process is extremely time and site-specific, and a more precise ability to predict the effects is needed. The pathway of the eroded material further down the catchment is often irregular and difficult to follow, so protective measures may be needed closer to the place where off-site damage is occurring while erosion continues.

Cultivation techniques are therefore very important. Mixed- and inter-cropping has the advantage that the soil is very rarely left completely bare. Zero tillage has obvious attractions, assuming that a layer of residues is left on the soil surface, and was shown to be beneficial in West Africa and elsewhere. However, there are reports from East Africa that cultivation produced better yields than zero tillage, while the soil structure was maintained. It seems likely that these differing reports are a typical case of the site-specificity of soils work, and a consequence of less erosive rainfall and different soil types in East compared to West Africa.

Soil Organic Matter (SOM)

Soil organic matter (SOM) occupies a crucial position with regard to soil fertility and its ability to grow crops. The microbial breakdown (mineralisation) of SOM provides nitrogen, phosphorus and sulphur as plant nutrients - in the absence of fertiliser or organic wastes, this is the only source. SOM can also increase the ability of the soil to hold other nutrients such as potassium. The structure of soil is also dependent upon SOM, which cements and stabilises soil aggregates and thereby makes the structure more porous and more stable.

The increased porosity is valuable in holding water in the soil. The presence of adequate SOM is associated with the level and activity of the soil biological population, which is essential for soil health. The ability to maintain an adequate level of SOM is therefore critical for soil as a medium for plant growth. Models are now available that can predict the future SOM level from the environmental conditions, the inputs of vegetable material and the soil type, but these still require improvement and testing.

In the last decade the behaviour of this reservoir of carbon in the SOM has become of heightened importance because it partly determines the carbon dioxide level in the atmosphere. World soils contain roughly 1500 Gt, with about 750 Gt in the atmosphere and about 550 Gt in the land biota (almost wholly vegetation), so the possibility of carbon storage in standing vegetation or in the soil is of great interest. However, the most immediate question is how much carbon is lost from deforested and newly cultivated soils, currently thought to be about 2 Gt per year. However, the most recent work suggests that tropical forests are in total net absorbers of about 0.9 Gt carbon per year. Much work on this subject is still needed.

Plant Water Relationship

No plants can grow without losing water, because of the need to take in carbon dioxide for growth through their stomata, though some plants have various mechanisms for water sparing or drought resistance. The weight of water transpired for unit weight of dry matter formed (the transpiration ratio) varies widely both with the plant species and the climate, but is usually of the order of 200-1000. This ratio is likely to be decreased by the steadily increasing level of carbon dioxide in the atmosphere, but the detailed effects are still not clear.

The total amount of water that can be transpired by a crop during its growth cycle is determined by the climate, the radiation interception, the canopy structure, and the internal physiology. If that amount of water is not available, as rain, irrigation or water stored in the profile, the yield is lowered. In rain-fed agriculture, the amount of available water stored in the profile is critical, and depends upon previous rainfall, the water-holding capacity of the soil and the partition between infiltration and run-off. The extent and ramification of the root system determines how much of this water in the soil profile can be utilised, and at what rate. The water not used in evapotranspiration or in run-off is then available for percolation and groundwater recharge - the second point of partition.

Water is also lost from the soil surface during plant growth, at a rate dependent upon the surface soil water content, and the rate at which radiant or adventive energy reaches the soil surface. Where the crop canopy is sparse, a large part of the total water may be lost in this way, and dryland crops often transpire less than half the rainfall. The use of mulch may improve this, but a thick mulch can retain rainfall, and allow it to be evaporated without reaching the plant roots.

The frequently poor utilisation of water by crops in dryland environments offers a number of possibilities for improvements. These include concentrating rainfall into a fraction of the land area to increase depth of percolation (a form of water harvesting), or use of intercropping to give a more complete or a longer-lasting canopy. However, there has to be a compromise between using the rainfall for immediate cropping, and allowing it to recharge the groundwater or maintain the river flow.

The total water use by a stand of plants on a given soil with given rainfall varies with the structure of the canopy, which determines the "surface roughness," and thereby the interaction between canopy and atmosphere. It also depends upon root penetration, so that the profile can be dried out to different depths. This is the cause of the general finding that woodlands transpire considerably more water than do short crops or grasslands. The replacement of one type of land use by another can therefore sharply alter the amount of water that can be left in the profile, or percolate down and recharge the groundwater or maintain river flow.

All these processes have been built into a number of models, based upon the single-dimension Soil-Vegetation-Atmosphere-Transport (SVAT) models and the Penman-Monteith equation, which has been extended to deal with incomplete canopies. At larger scales these are coupled to above- and below-ground water transport models (see below), incorporating variations in soil type, vegetation structure and topography if necessary. These are often large and complex, but are becoming steadily more accurate. In the limit such models are being built into the Global Circulation Models used for climate prediction. The theoretical underpinning is therefore strong, with the weakest part probably in the simulation of the distribution and function of root systems in relation to soil properties.

One of the most difficult situations to research and to understand is that with controlled competition between two or more species, as in intercropping or agroforestry. The mechanism of competition may be for light, nutrients or water, of which the last is likely to be important in all systems in which there is a net water deficit. The measurement of how and to what effect the different species and plant individuals divide the limited available water supply is technically complex, but obtaining a net benefit from having two or more species present depends upon this partition process. Most research in more-developed countries has focused on the monoculture situation, even in work on forests.

Hydrological Processes

Climate is defined by the annual means, the distribution of the variables according to season, and the between-year variability. All are critical for rainfall, because this determines the amount of precipitation, its intensity and the frequency of extreme events. These affect infiltration, runoff and erosion, horizontal and lateral transport and storage in soil profile, drainage and groundwater recharge, and associated questions of water quality. Groundwater is a most important component, and its lateral flow towards the rivers helps to maintain their volume. Lateral flow of groundwater may cause transfer from one watershed to another, but it is unlikely to be a major factor. The underlying theory of water transport in soil is strong, but the accurate measurement of the actual soil parameters in the field situation can be difficult.

The basic understanding of the movement of fertilisers, pollutants and other solutes is good, but work in the field is difficult because of the heterogeneity of soil structure. Strongly structured soils often have a bimodal porosity, which means that during rapid percolation, when large pores are waterfilled, the rate of penetration of solutes can be much faster than expected. When rivers are contaminated by such chemicals, the water quality will fairly soon return to normal if the sources can be controlled. Contamination of groundwater is potentially far more serious, in that groundwater may take decades or centuries to be replaced.

Watershed research is heavily dependent upon the application of computers and mathematical modelling, which are now used in data storage, real-time control systems for river management, simulation models for pollution of rivers, and large models of subsurface flow, groundwater behaviour and total watershed hydrology. The SHE (Système Hydrologique Européenne) is an example of such large and data-hungry models for predicting water movement over considerable areas.

The use of modelling in controlling large irrigation schemes is one aspect of this type of work, which is made more difficult because high salt and sodium contents can change the hydraulic properties of a soil markedly. Other uses are in predicting the water relationships in watersheds under different forms of land management. There is a large hydrological modelling programme for the whole Mississippi basin operated by a part of the World Climate Research Programme. There are also now programmes that aim to connect hydrological models to other spatially arranged models, such as the modelling system for hydrology, farm economics, land use, ecology and water quality that has been constructed for two river basins in the UK.

Problems in Terrain

There are four possible situations that might require research, but which all merge into each other without sharp distinctions. These need to be kept in mind in considering soils problems. First, there is the soil under cultivation without immediate serious problems but with mediocre productivity, and which is likely to be degraded over time. The task is to enhance productivity with technologies that together form a sustainable agriculture. The success depends greatly upon the inherent resilience and stability of different soil types; Greenland et al. suggests that only 10% of tropical soils have the natural fertility to be suitable for continuous cultivation without enhancement. This has been a CGIAR task for many years, and it has developed considerable experience.

Second, there is the soil which has already been seriously damaged or degraded by poor management, and which needs reclamation or rehabilitation before it can become productive. The most common causes are probably excessive shortening of fallow periods in shifting cultivation, and loss of soil by water or wind erosion due to poor cultivation and conservation techniques on slopes. Productivity will have sharply declined.

Third, there is the marginal soil, which should probably not be taken into cultivation at all because of its infertility or susceptibility to damage, but is progressively more likely to be used because of population pressure. This will include very steep slopes, very sandy soils of poor structure, and soils in areas of little or irregular rainfall. Some of their problems will be similar to the degraded soils, but rehabilitation is unlikely to be cost-effective.

Fourth, there is still some potentially good unused land which is gradually being taken into cultivation, and the task is to ensure that this is done in a careful and efficient way so that its productivity is maximised. This would apply to some land in long-term fallow with shifting cultivation, some forest and some savanna land, much being in South America. However, even there, the new land very often has constraints that make it expensive to bring into cultivation, and make intensification of existing farmland more attractive.

a) *Steeplands*. There are many examples of grave problems caused by population increase and movement onto steeplands. The classic case is N-W China, with the heavy erosion of the loess soils. Even more critical cases are now found in Nepal and the Philippines and in the Andes and Central America. Where these steep soils have not been cultivated before, and where the farmers may be incomers, there may be little or no local traditional guidance for farmers. Extension advice is rare, and may be mistaken, again because of lack of experience of these problems. As land pressure increases the fallow period shortens and the cropping period lengthens, so that the net landcover fraction decreases steadily. Ultimately good cover does not regenerate. Where systems for mulching with branches from forest trees are in use, the gradual removal of forest makes the system unsustainable.

The standard way to use such land is by terracing or bunding. This may not be used because returns on the investment of labour and materials are perceived to be insufficient or too uncertain, or because of lack of knowledge, lack of organisation or funding, or simply because the speed of the degradative process overwhelms the farmers. In many cases the land is simply not appropriate for arable farming, and should be left in forest, or converted to tree crops.

Uplands are typically a major source of river silt if their tree cover is removed, due to logging, harvesting of fuelwood, or for agricultural use. The consequence is that dams are rapidly made useless, and that heavy flooding and silt deposition occur.

(b) (i) *Rolling land, low rainfall.* The low average rainfall, high potential evapotranspiration and high variance of climatic parameters makes agriculture in this zone unavoidably risky without irrigation. The Report on CGIAR commitments to West Africa considered that the impact of research in this type of area had been weak. However, sustainable agriculture in this situation depends upon maximising the use of all water, and there are techniques for water harvesting or collection of runoff from occasional storms.

The prospects of producing cultivars that are still higher yielding in droughts than present ones by breeding or genetic engineering need to be assessed. If the risk of frequent and damaging drought can be reduced, the farmer may find it worthwhile to invest in agronomic inputs. In dryland farming, soil fertility and water supply have to be fully integrated to maximise production.

(ii) Rolling land, good rainfall. These areas are core agricultural parts of a watershed. The rainfall can encourage more intensified agronomy to give higher yields, but can also cause higher leaching of nutrients and agrochemicals. Perennial tree crops are an option with advantages for soil protection. The problems are the central ones of overcropping, too little fallow, and too few inputs. The use of pasture as an alternative to bush fallow may be advantageous in some areas where there is livestock.

(c) (i) Flat land, low rainfall. The position of this terrain in the watershed should make irrigation a practical option, as in much of South-east and South Asia. The problems are poor distribution of water, salinity of soils and water, and loss of structure if sodium is a major exchangeable cation. A particular problem arises if there is dependency upon groundwater, and this has become saline.

In principle there are biophysical solutions to most of these problems, though much adaptive research is always essential. The cost or the limited managerial capacity may, however, make these solutions impossible to apply. Some of the problems are the off-site result of actions higher up the watershed, in particular the siltation of dams and watercourses. Generally the management of irrigation systems leaves much to be desired, and this is a major cause of low yields. There have been encouraging improvements recently, usually associated with giving a greater role to the farmers themselves.

The problems are therefore socioeconomic and environmental rather than strictly biophysical. Where the Green Revolution has been successful, there are now growing problems of contamination of groundwater with nutrients and pesticides. The solution for nutrients must lie in better control over the use of fertilisers, manure, intensive animal production units and other agricultural sources. For pesticides, much can be done by using only minimum inputs, but more fundamental solutions are to use Integrated Pest Management, with biological control measures.

(ii) *Flat land, good rainfall.* Irrigation is probably not necessary in such areas, but there may be low insolation problems due to the cloud cover. The natural vegetation is probably forest, and tree crops are a valuable method of land use, providing varying degrees of protection to the soil. The heavy rainfall is likely to dilute agricultural pollutants, so that fewer problems are likely than in drier areas.

(d) *Wetlands.* These occur at most levels of a watershed, if the topography is appropriate, but most often in the lower river valleys or deltas. They often are the habitat for important wildlife, and are the subject of the international Ramsar Convention on conservation. Agriculture will again be largely wetland rice cultivation, and problems will be largely off-site ones from higher up in the catchment. This zone is well adapted to aquaculture; if this is intensive, it may itself become a source of nutrient pollution. If much irrigation is practised higher up the river, and the river water becomes saline, it is in this zone that the worst off-site effects are felt.

Benefits of Management

The above has assessed the problems of an idealised watershed in terms of topographical position. A main objective of research and extension is to increase yields, or at the very least to prevent declines such as have been recorded in Africa. Prevention of soil loss can be done by a number of well-known techniques, which need selection of the most appropriate, and adaptation to local conditions. The impacts of erosion are felt on-site, both where soil is lost and where it is deposited, and in other land use units where silt is deposited, on fields, in watercourses and in dams. Apart from major interventions such as dams and channel alterations, water flow can be modified by altering the rate at which water reaches the river and affects the hydrograph. This can be done by altering land use (for example, from forestry to grassland) or by changing the "first partition point," between infiltration and runoff by altering the soil surface properties with cultivation, mulches, SOM level, water ponding or others.

Water quality can be controlled by minimising the use of fertilisers and agrochemicals, controlling feed lots, stopping large inputs of organic material to watercourses, preventing accidental spillages and managing irrigation systems to control saline drainage water. In fact, water quality and quantity are closely linked. In a simple case, the concentration peak that travels downstream in a river after a pollution incident will be dispersed at a rate that depends upon the volume of water, its flow pattern and the river's course.

All these processes must therefore be seen as a network of interactions. This complexity is the main challenge of integrated watershed work, and any intervention with a single objective raises the danger of unforeseen problems. Sometimes the results are totally unexpected, as with the damage suffered by Mediterranean fisheries when the silt load of the Nile was reduced by the building of the Aswan dam.

This idealised watershed will almost certainly have cities and towns. These will all have demands for water of acceptable purity, for domestic and industrial use. The waste systems will also produce sewage, which has to be purified, and the effluent fed back into the river - these are potential point sources of heavy pollution. They are probably even more likely than agricultural pollution to cause eutrophication, excessive algal growth and possible anaerobiosis in the rivers and lakes. The management of surface waters in these cases is itself a difficult and professionally challenging task. Furthermore, the cities will contain soils, and the management and use of urban soils is a growing issue.

NRM Research

The value and applications of NRM research on a world scale are incontrovertible. They range from making land agriculturally useful by detecting specific deficiencies, for example copper in Australian soils, to the understanding of hydrology and soil physics that allow irrigation schemes to be designed. In agricultural terms, NRM and production research overlap, and often fuse. This research over decades has built up a considerable knowledge bank on soil management, erosion control, river management and groundwater use, and much applied and adaptive research in the less-developed world has drawn directly on this. In some important cases this adaptation has been extremely successful, as is represented by the Green Revolution, in which improvements in germplasm, plant nutrition, plant protection and water supply all had their part.

This is quite appropriate for strategic research, because it is intended, by definition, to produce knowledge of the underpinning or cutting-edge type, rather than immediate solutions. It is more a cause for concern if applied, or especially if adaptive research is not used. However, there do appear to be large areas where NRM research results have simply not been used. It is usual to blame this on the proposition that the proposed solution is socioeconomically difficult to apply, or simply unprofitable, or that conditions in these areas are so different from those elsewhere that the knowledge bank cannot be applied, or that the solutions it provides prove unreliable or partial.

Where the socioeconomic conditions are favourable, so that farmers can get a sufficient cash return for their work, there are successes, such as the well known Machakos work. A number of small-scale pump- or gravity-fed irrigation schemes in Mali and Kenya have also been successful. These depended variously on sound characterisation of the environment, and on participatory approaches with farmers and local authorities. The watershed work in East Africa in the 1950s and 1960s provided essential underpinning information concerning land uses, such as that, in particular locations, indigenous forest could be replaced by softwood or tea plantations without affecting the water yield of the watersheds, or damaging the soil. The level of erosion protection that was necessary for cultivation, and the acceptable level of grazing, was specified. Where these results were applied and followed, they were successful, but elsewhere they were ignored.

A recent update on the situation in Kenya shows how successful the catchment approach has been when consistently applied in a fully participatory way. This emphasises how often major programmes in soil conservation have not been followed through, and have been allowed to deteriorate when the farmers who should benefit are not convinced of the value of the work and of their own role. It is clear that sound technology and biophysical analysis can be successfully applied when the social conditions are right. As an example of this success, the imports of maize to the Machakos district have decreased by more than half on a per capita basis over the period since 1945 during which the population has increased threefold. .

An excellent example of a "catchment" problem at a totally different scale of the whole of Egypt. The uses of land and water resources are inextricable interrelated, and most of the agricultural problems relate to water quantity and quality, and the overall sustainability of the system is dependent on these. Scientific water management is here essential. There are clearly many situations where existing strategic knowledge can be and has been used for applied and adaptive research, and has produced biophysically successful results. Much of the work in applying this basic information can probably be described as adaptive research. However, it is rarely a question of going into the field and making measurements according to a standard technique and applying solutions from a textbook. Professional scientific understanding, and the ability to modify procedures and technologies according to the properties of a particular area, are essential for this type of work.

Where existing methods have not been applied it is logical to call for simpler, cheaper and easier methods, and research should certainly be directed towards discovering these. However, there is a point beyond which it is unreasonable to expect cheaper and easier solutions, if the biophysical situation is in fact technically difficult and complex. If better or cheaper biophysical solutions are not readily provided, it may be possible to use socioeconomic interventions that make the existing lowest-cost biophysical solution acceptable.

If much-superior biophysical solutions are not available, policy interventions that modify the socioeconomic circumstances may still be possible, so that modest low-cost biophysical solutions become more acceptable. As is clear from 4.5, biophysical research alone can rarely solve problems in the real world, but it is very difficult to solve any of them if the biophysical basis is not available and correct.

From recent audits of a set of 335 agriculture and forestry projects, 8% had adverse results involving soil erosion or salinisation. Of the various adverse results, shortfalls in performance were attributed to one or more of inter alia inadequate knowledge of: physical conditions (42%); techniques (28%) and; social institutions (12%). The stakes for developmental interventions are large indeed, and the returns to improving knowledge of NRM phenomena are likely to be great.

Natural Resource Management Issues

Natural resource management issues are primarily national and local ones. They can usefully be studied comparatively and cross-nationally, but they are invariably quite political, or have political implications. Thus, solutions need to be developed quite inductively and in accord with national and local traditions, cultures and values, not to mention their respective and often unique biophysical conditions.

This said, international institutions can play a valuable role in this area, affecting the climate of opinion through their research and publications, and improving analysis through the development of concepts, methods and measures. The first is probably the most important, though it is hardest to plan or to prove.

In the area of irrigation water management, for example, a significant movement has been evident toward new, more participatory, more efficient and more effective management systems over the past 20 years. This was initiated with Ford Foundation support in the Philippines, India, Nepal and Indonesia; it was supported by donors, first USAID in Sri Lanka, Indonesia and

Nepal, and then the World Bank in the Philippines and elsewhere.

Over time, decision-makers and state sector managers in less-developed countries have confronted a growing body of evidence and analysis that pointed the way to more effective and beneficial irrigation management. New assumptions have come to be widely shared about the feasibility and desirability of adopting less bureaucratic approaches.

With regard to watershed management, similar in this respect to forest management, there is a long tradition of technocratic perspectives and control. Indeed, the concept and strategy of bureaucratic reorientation was prompted by experience with forest management agencies that was similar to what was observed with irrigation departments in Asian countries. A similar progression, though slower, has also been apparent in the forestry area, as forest agencies in many places have come to recognise that the resources they are responsible for can be more beneficially and effectively managed in cooperation with persons living in and around forested areas.

Watershed management is now receiving similar attention, and lessons from other NRM areas are being brought into this domain mutatis mutandis—and sometimes not in this way, being simply transferred from other domains, which is a prescription for difficulties, since different areas as well as different resources require management systems that are adapted to the particular biophysical and socioeconomic circumstances. There is more interest now in participatory approaches than there was some ten years ago, and this reflects in part the actions of international institutions and researchers who have raised issues and examined alternatives.

Certain important issues in NRM such as land tenure can be identified as candidates that international institutions should study and provide advice on. But it is hard to say that these can be "best" addressed at the international level unless it is clear that they will not be (adequately) tackled at national and sub-national levels, so the international level only becomes the best by default. In fact, the research community is international, and every encouragement should be given to researchers around the world, in their respective national and local settings, to try to illuminate important issues such as land tenure and its impact on sustainable NRM. But this is not necessarily something done "at the international level."

Resource Management Issues

Resource management issues naturally differ from country to country but, even where they are similar, performance varies greatly because of significant institutional differences. Attempts to generalise about successes or otherwise are thus fraught with difficulties, but must be addressed to help to define roles for the various potential actors, local, national, and international. Needless to say, there are many actors who have explored their possible roles, although the deliberations are not always documented in the public domain. These include planning documents for concerned NGOs, institutional development documents for governments considering, for instance, how best to set up environmental protection agencies, and so on.

One of the players, until recently a somewhat coy one in this area, is the World Bank. It was long and persuasively argued that the Bank had neither the mandate to intervene in matters so deeply rooted in national culture and policies, nor the technical and financial clout to make sufficient a difference. But all this changed radically in the late 1980s as signalled, for instance, by the creation of its Environment Department and counterpart regional environmental operational units. In the process of entering upon these new institutional arrangements, the Bank took various steps to consider its position in resource-management related lending, which has now become strongly oriented to environmental interventions.

In 1991, the Government of the Indian state of Rajasthan created a new multidisciplinary department of Watershed Development and Soil Conservation, building on the Soil Conservation wing of the Agriculture Department and drawing in staff then from other departments. The programme had substantial funding available from the central government's National Watershed Development Programme for Rainfed Areas and a World Bank-funded Integrated Watershed Development Project.

Within two years, it was carrying out conservation and development work on over 100,000 ha in over 250 locations. It emphasised local participation through User Committees that were facilitated by department staff. Rules and procedures for snaring costs and benefits among local residents, and between them and the government agency, were worked out consultatively, with different formulas adopted in different locations. In both 1993 and 1994, the programme was awarded a Certificate of Merit from the National Productivity Council of the Government of India for its development of appropriate technology and for its high degree of community participation.

A very decentralised approach was taken, with the delimitation of 250 watersheds averaging about 4,000 ha each. A multi-disciplinary (multi-departmental) team was assigned to each watershed, to work with the villagers residing therein. Great encouragement was given to be experimental and to innovate, with the result that a visiting GOI official spoke of "a widespread unleashing of the creative talent of field staff.". The programme expanded rapidly, to cover ten times the area previously served by soil conservation schemes, and this was possible with staff reorientation and enthusiasm, and with village response.

Krishna considers as the main factors in the rapid success: (a) the experimental and adaptive approach taken to technological innovation, (b) the rapidly and strongly enforced administrative coordination, and social organisation among government agencies, and (c) the attention and support given to social organisation, i.e., to user and community participation. Whether the programme will retain its operational and philosophic thrust without his leadership is something yet to be assessed. But this case has shown what even a bureaucracy known to be rather lethargic and a set of communities known to be quite traditional and isolated can achieve in short order.

In 1990, the International Irrigation Management Institute was contracted by USAID to design a new kind of NRM project. USAID's previous support of a water management project focused on the Sri Lanka's Gal Oya irrigation system had shown good results by engaging water users in a multi-tiered structure of organisation, starting at the field channel level and extending upwards. The largest and one of the most run-down irrigation systems in the country had become one of the best managed within a few years, doubling the efficiency of water use.

After extending this participatory irrigation approach to four more major irrigation schemes through a follow-on project, USAID and IIMI were interested in tackling the resource management problems of whole watersheds in a participatory manner, to see how far the methods developed inductively, i.e., in a learning process mode, in Gal Oya could be extended beyond irrigation.

Two whole watersheds were chosen as pilots for the Shared Control of Natural Resources, known as SCOR: Huruluwewa in the North Central Province, and Nilwala in the Southern Province. Each had a major irrigation system downstream that would be jeopardised in the long run if shifting cultivation continued in the upper catchment area and accelerated the deforestation and erosion already evident there. Sub-watersheds in the range of 75 to 600 ha were identified (similar in scale to those in Rajasthan). The project was not implemented by a single department as in Rajasthan. Rather, IIMI recruited interdisciplinary teams of professionals with a variety of skills and backgrounds, but all committed to a participatory approach.

In Sri Lanka, SCOR project combined participatory assessments of present land and water use patterns, capabilities of resource user groups and support services, socioeconomic status, status of resource degradation, and potential for development, with more sophisticated geographic information system (GIS) technology. Persons were recruited, trained and deployed in the field as "catalysts" (as in Gal Oya) to work with communities to promote social organisation. Here they worked not just with irrigation water users but also with upstream resource users.

To assess progress and success, the project is monitoring sediment concentration and sediment load in the streams and rivers of the respective watersheds, changes in soil fertility and soil loss, biomass, water quality, and rainfall runoff and infiltration; water use efficiency in agricultural activities, factor productivity and profitability, cropping intensity, and cost reduction/value added; and effects of land (including common property) covered by group activities in terms of production and protection, value of investments made by user groups, number and type of commercial activities undertaken by groups, number of policy and procedure changes associated with project and user activities, and returns to shared control of land and water resources. The strategy is characterised as seeking an effective balance and blend of technologies, organisations and resources.

This is quite a dramatic step forward toward integrated management of soil and water resources. While it is operating on a smaller scale than the Rajasthan case, it is similarly interesting because of its promotion of commercial enterprises that are intended to give greater value to natural resources in a way that should create incentives for their conservation, and because it is linking upstream and downstream resource users into a common forum to assess and decide on all forms of NRM. Implementation is involving all relevant government agencies and the several levels of local government that have jurisdiction within the watersheds, so that the new people's organisations will have understanding and support from public-sector agencies.

Beginning in 1989, staff of Save the Children, an international NGO, began meeting with landowners in 22 small, multi-owner watersheds (averaging 9 ha in area) in the central plateau of this impoverished country. The region was quite hilly (average 12 percent slope), and soil erosion was contributing to steadily declining agricultural yields. Although Maissade Watersheds, Haiti is not known for collective action, a majority of the landowners were members of informal, self-organised and self-governing farmer co-operatives known as groupman (with 8 members on average). The average size of holding was 2.5 ha, spread in usually three separate parcels. A survey showed that the average watershed in the area had 9 agricultural parcels with 9 different landowners.

Transboundary erosion was common in these watersheds and could hardly be avoided where the soil infiltration rate was low and there were few if any soil conservation structures. All but the topmost farmers both lost and received soil through erosion, creating a perverse kind of interdependence. Downhill farmers in principle could benefit, but the rapid runoff itself caused problems, and the spread of gully erosion as well as sheet erosion was a threat to most farmers on these hillsides.

State efforts to curb erosion had proved ineffective, according to Murray. Yet this NGO programme proved quite successful in mobilising local voluntary labour to construct checkdams, working with rural residents on the basis of informal cooperative action that they were familiar with (groupman). Within two years, 10 of the principal ravines had been treated, with partial treatment of another seven, while five remained without much soil conservation accomplished.

A total of 590 checkdams were constructed by the groups, averaging 27 dams per watershed, which were estimated to retain an average of 39 t/ha/y of soil. The amount of labour mobilised on a voluntary basis for soil conservation was impressive: 32 person-days per year from landowners. Perhaps more impressive was the voluntary contribution of 18 person-days per year from persons who did not own land in the watershed. Building on traditions of cooperation and mutual self-help, there was a substantial community effort to counter the effects of erosion that were visibly eating up future production possibilities.

The Haiti case is considered a success not because of the large scale of operation as in Rajasthan, or the innovative institutional development as in Sri Lanka, but because (a) it operated under very adverse sociopolitical conditions and in a physical environment where soil conservation and watershed protection was so urgently needed, and (b) it showed possibilities for collective action that had hardly been considered possible by most social scientists and most policy-makers. Extremely poor and isolated rural people were willing to invest labour and organisational effort to preserve as best they could their agricultural possibilities for the future.

6

Soil Conservation

Soil conservation in its widest sense includes not only control over erosion but all those measures like correction of soil defects, application of manures and fertilisers, proper crop rotations, irrigation, drainage etc. which aim at maintaining the productivity of the soil at a high level. In this sense, soil conservation is closely allied to improvement of land use in general. Large areas in all parts of the country have been rendered useless as a result of soil erosion and areas which suffer from moderate or slight erosion and whose productivity is reduced as a result of soil losses are very much larger still.

Sheet erosion, which consists in the washing away of the fertile top layers of the soil, is the most extensive form of erosion, occurring even on moderately sloping lands. It causes enormous losses to agriculture every year by reducing the productive capacity of lands. Gully erosion, which generally starts after sheet erosion has remained unchecked for some time, has already rendered large areas useless, and is steadily increasing. In the dry western part of the country, erosion as a result of wind action and covering of croplands by desert sands along the margins of the Rajasthan Desert constitute serious problems.

The most important cause of erosion is destruction of forests and other vegetation from sloping lands, desert margins and other areas susceptible to erosion. Vegetation acts as a protective cover against the forces of wind and water, protecting the soil from being washed or blown away and preserving the physical and-hydrographic balance of nature. Forests-for instance, provide the most effective protection against erosion on hill slopes. They break the force of run-off by impeding the flow of rainwater down the slopes and .by absorbing large quantities of it in their dense mat of undergrowth.

This absorbed water flows away slowly over a period of time; a large part goes into the soil, flows undergrounds, feeds springs and streams and is available for utilisation in the foothills and plains. In this way, the hill slopes are protected from erosion, the flow of streams is regulated, the danger of floods is reduced and sufficient quantities of water are available in dry periods. But, when the protective cover of forests is destroyed, this natural balance is disturbed. Rainwater flows down the slopes unimpeded at great speed and carries with it large quantities of soil and other loose material.

The hill-slopes are denude of valuable soil and lands' in the foot-hill zone where this unassorted mass of sands and gravels is deposited are in turn rendered unproductive. Most of the water flows away during the rainy periods with the result that on the one hand floods are more frequent and more severe and, on the other, little water is available during the dry periods. Ground-water, supplies are also reduced as much less water is absorbed in the soil than before.

Destruction of trees and natural grasses in dry areas has similar harmful effects. Trees act as wind-breaks, reducing the force of the wind, and the grasses bind the sandy soils. But, when such protective vegetation is destroyed, the sandy soils, exposed to the full force of the wind, begin to be blown away. Large areas in the marginal zones of deserts are thus rendered unproductive by the deposition of sand. It is believed that many deserts of the world (including the Thar Desert of India) have increased in area in historical times by this process.

The causes of destruction of forests and soil erosion, and the nature and severity of the erosion problem vary greatly in different parts of the country. In some areas as in the forests of Assam, Bihar, Orissa and Madhya Pradesh, shifting cultivation, which is practised by the tribal people living in these areas, is a major cause of destruction of forests. Unregulated grazing is the cause of destruction of forests and consequent soil erosion over large areas in all pares of the country.

In the north-western Himalayas, for instance, grazing by cattle, sheep and goats is the most important cause of depletion of the vegetation cover and soil erosion. Similarly, over large parts of peninsular India, the destruction of forests and soil erosion are due mainly to over-grazing. Intensive felling to obtain supplies of fuel or timber, and clearance of forests for extension of cultivation Under the pressure of demand for agricultural land from the increasing population are other important causes of deforestation and soil erosion.

Besides the damage caused as a result of the destruction of forests, considerable erosion results from faulty land use practices on farmlands themselves. Failure to practise such measures as ploughing along the contours on sloping lands, proper crop rotations and in particular growing of cover crops are causes of erosion over large areas. Much damage originates also in fallows, grazing lands or uncultivated waste lands which are generally neglected.

Measures for controlling erosion and restoring the productivity of eroded lands can be divided into four classes:

— *Regulation of land use*—This includes all measures for securing such alterations in the existing patterns of land use as are necessary to ensure that the different types of lands are used according to their land use capability i.e. the use for which, in view of their physical characteristics, they are best fitted. Retiring cultivated lands in highly erodable areas from cultivation and bringing them under forests or grazing ; restrictions on or closure of grazing in badly eroded forests or grazing lands, and settlement of shifting cultivators to permanent cultivation are some examples of the types of alterations which are needed.

— *Afforestation and preservation of forests* by scientific forest management.

— *Improvement of land use practices on farm lands.* This includes such measures as ploughing along the contours and strip-cropping on sloping lands; proper crop rotations ; application of adequate manures and fertilisers; care of fallows and other uncultivated lands.

— *Engineering measures*—Under this are included construction of bunds and terraces, check dams, channels for drainage of surplus water, gully plugging, etc.

A comprehensive programme of soil conservation for an area would include all four types of measures, although the relative importance of the different measures would vary greatly in different areas depending upon the particular conditions of the area.

As a large part of the sail conservation work has to be done by the farmers, proper understanding on their part of the nature of the erosion problem, and their active participation in soil conservation programmes are essential for the success of these programmes. Improvements in farming practices depend entirely upon the farmers. Government's function is mainly one of convincing them of the need for such improvements and demonstrating the correct methods of adopting them.

Financial assistance, in the shape of supplies at reduced rates or in other forms may also be given. Engineering measures have to be taken mostly on the farmers' fields. These may be taken by the farmers, individually or in cooperation, with technical or financial assistance from the Government; or the work may be done by Government and the cost (or a part thereof) recovered from the farmers. Finally, restrictions on usage in forests can be really effective only if ths farmers, graziers and other users of forests understand the importance of these and feel that they are essential in their own long-term interest, besides being vital for the welfare of large populations in the plains.

Education for soil conservation, publicity and demonstration aimed at creating among the general public and especially among the farmers an awareness of the erosion problem, its causes and effects, and what they can do to control it must form a very important part of soil conservation programmes. Formation of associations of farmers for soil conservation work has also been proposed in order to provide a suitable medium through which soil conservation measures can be taken on a cooperative basis at the village level.

Steps for the control of erosion and conservation of soil have been taken for a number of years in certain States like the Punjab (afforestation in the Sivalik Hills) and Bombay (bunding and terracing work in the Deccan). More recently, soil conservation work has been taken up in several other States also. But there has been no country-wide effort in this direction so far, and even in States where the work has been going on, this has been on a very limited scale. The programme for soil conservation in the Plan, though small in comparison with the magnitude of the problem, marks the beginning of a. country-wide effort to tackle it.

There are many limitations to undertaking a larger programme at this stage. Very little work has been done on soil conservation so far ; data on such basic items as soil characteristics and type and severity of erosion in different parts of the country is lacking, and technical personnel with the necessary training and experience is limited and has to be drawn from many different fields. These limitations will be largely overcome during the period of the Plan. The necessary administrative machinery will be set up at the Central and State levels; survey and research organisations will be established and essential data collected ; suitable legislation enacted, and a much greater consciousness of the erosion problem will be created. As a result, it will be possible to take up a more adequate programme in subsequent years.

Programmes for soil conservation and improvement of land use during the period of the Plan should be worked out for each State by its Land Utilisation and Soil Conservation Board, the formation of which we are recommending. These programmes should be based on an assessment of the erosion problem in the State. Such an assessment can best be made by a rapid survey of the reconnaissance type by.which the major erosion affected areas are demarcated and the types and degree of erosion in each area broadly indicated.

One or more areas of suitable size should be selected for work during the period of the Plan. Preference should be given in making this selection to areas which are representative of much larger region? suffering from erosion so that the experience gained in these would be applicable to the largei areas. In States where soil conservation work is already going on, the possibility of extending the scope of this work should be carefully examined. The State Plais would be reviewed and approved by the Central Land Utilisation and Soil Conservation Organisation. Assistance from the experts of the Central Organisation may be obtained by States which may need them in drawing up these programmes or subsequently in their execution.

In 1981, the Ministry of Agriculture and the University of Bern (with the support of the Swiss Government) initiated the Soil Conservation Research Project (SCRP), which generated one of the first systematic data sets on the magnitude and the severity of erosion. In 1983, the Ethiopian Highland Reclamation Study (EHRS) was carried out by national and international experts, which reviewed the main reasons for drought and proposed for Conservation Based Strategy to address it.

Awareness has also led to action and the Government of Ethiopia (supported by various donors, international agencies and NGOs), has made largescale investment in soil conservation and land rehabilitation measures. The rehabilitation of degraded lands, which started through food-for-work relief assistance following the 1974-1975 famine, has become a major component of the Government's approach to mitigate the impact of soil degradation in many regions of Ethiopia.

This approach has focused on a) soil and water conservation; b) construction of terraces, check dams, cut-off drains and micro-basins, and c) afforestation and revegetation of fragile and hillside areas. The focus was on building physical structures to control soil erosion and to rehabilitate degraded lands and massive efforts were undertaken in this regard. This effort has resulted in many ecological benefits such as restoring farmlands, increasing soil depth, water holding capacity and improved woodlot and pastureland.

As important as these ecological benefits were, the large-scale soil conservation efforts of the 1970s and 1980s had some serious shortcomings. First, these structural conservation measures were found to be too costly. After all the investment, not more than 10 percent of the cultivated land has been covered. For example, it was observed that the labour input required for constructing fanaya juu bunds is ten times more than planting grass strips, which are reasonably effective in reducing soil loss and increasing moisture content and water infiltration.

Second, farmers were reluctant to adopt such labour intensive measures (without getting tangible benefits in terms of food or income). Third, there was little systematic effort made to incorporate indigenous soil and water techniques, and not to consider the loss of farmland for conservation. Finally, there is no obvious relationship between this large investment in land rehabilitation on one hand and improvement in the food security and income of farmers on the other. This has also been observed by several other studies.

It is also widely acknowledged that the soil conservation policies and activities of the past decades have not been successful. It is fair to say that the technologies focused narrowly on arresting soil erosion without fully considering the underlying causes of low soil productivity, socio-economic factors, and the need for tangible benefits to be attractive to poor farmers. Indeed the evaluation of the Swiss- supported SCRP has revealed that the project focused on the technical aspect of soil erosion and paid insufficient attention to other aspects of land management, including the role of livestock (soil compaction), traditional ploughing methods, root system, land-use dynamics and policy and socioeconomic issues.

Another major weakness of the land conservation and rehabilitation effort over the past twenty years has been its top-down approach and that it has not shown demonstrable change in the day-to-day lives of farmers in terms of improving food security or income. This was also true of the Conservation Based Development recommended by the Ethiopian Highland Reclamation Study, which sees conservation measures (that has long-term pay off) as an important step in the attempt to increase agricultural productivity among small-scale farmers. This approach underestimated the short-term cost of some of the conservation measures (labour intensity and giving up of land used for crops or grazing), which were equally unacceptable to many of the poor farmers.

In the late 1980s the weakness of the top-down approach in planning and implementation was recognised by the previous Government (Derge) and initiated the development of the National Conservation Strategy to provide legislative framework for natural resources management. Nevertheless, this did not come into operation, as the Government was fully absorbed in internal civil conflict that led to its collapse in 1991. The Transitional Government of Ethiopia (TGE) that replaced the Derge recognised, in principle, the need to shift from the top-down approach to soil conservation and include the participation of farmers in the planning and implementation process.

Improving Soil Fertility

The role of fertiliser in improving the declining nutritional status and productivity of Ethiopia's soils is widely recognised. Fertiliser use has increased from 947 metric tonnes in 1971 (when it was first introduced) to 1,42,000 metric tonnes in 1992. Fertiliser use has increased from 246 722 metric tonnes in 1995 to almost 300,000 in 2000. This is partly due to the decision of the TGE allowing farmers to buy fertiliser with 100 percent credit in 1995. The Government's National Extension Package programme that put heavy emphasis on accelerating production using external inputs, often suited to higher potential areas, has also contributed to such a dramatic increase.

Production has more than doubled, mainly in maize, due to its responsiveness to fertiliser. But the price for maize plummeted to a record low level (by about 80 percent in surplus areas) and many farmers faced serious difficulty in paying back the credit they took to buy fertiliser, which is a major disincentive in using fertiliser and growing maize. For example, the price of Urea fertiliser in one of the surplus areas by 2002 was six times more than the price of maize while DAP was nine times more, making fertiliser effectively out of the reach of smallholders.

In Ethiopia, Phosphorus (P) and Nitrogen (N) were the most crucial limiting factor to plant growth, and not Potassium. As a result, a blanket recommendation of 100 kg DAPS and 50 kg of Urea were formulated for Ethiopia. The Sasakawa Global-2000 through its demonstration trials have shown that 100 kg of DAP plus 100 kg of Urea are sufficient for most crops except teff and maize which would require 50 kg of DAP and 200 kg of Urea. However, farmers do not often go along with the recommended practices but follow practices they can afford (often half the recommended rate).

It is now widely recognised among experts and policymakers that the increasing application of fertiliser at the current price will not be affordable to many farmers and possibly the Government (that lacks foreign exchange to import fertiliser with out outside support). Thus, extension and research should accord a high priority to find an economically viable option that uses fertiliser in combination with other local available organic sources.

Recognising that resource-poor farmers do not adopt recommended packages at once, farm trials have been conducted in semi-arid areas to show the relative contribution of agronomic practices and fertiliser on maize yields. The result revealed that raw planting alone (no fertiliser, late weeding, and flat planting) resulted in a yield increase of 37 percent from the controlled plot and if early weeding and tied ridges are used yield increased to 73 percent from the control plot. The findings underline that the most desirable and effective option in increasing yields significantly is the combination of using both fertiliser and agronomic practices.

It also underscores the critical role of good agronomic practices since the application of fertiliser alone (without good agronomic practice) has resulted in lower yields than plots using good agronomic practices without fertiliser. Another study has also shown that tied ridges (very efficient in capturing moisture) in dry land areas have increased yields in crops such as sorghum and maize by 50 percent to 100 percent. However, farmers do not like tied ridges because they have to be built by hand which is highly time-consuming.

The soil fertility initiative (SFI) for sub-Saharan Africa was initiated in 1996 at the World Food Summit (WFS) under the coordinating role of the World Bank to foster global partnership of donors, African governments, NGOs, development agencies, national and international research centres to restore declining soil fertility in SSA. The SFI could have the potential to enhance soil productivity and promote the intensification of agriculture through a combination of high and low input technologies among small-scale farmers in Ethiopia, if it were implemented through the appropriate Ministry and in a participatory way.

The National Fertiliser Industry Agency (NFIA), which was established in 1994 (formerly a unit under the MoA), emerged to take a lead responsibility in introducing the SFI in Ethiopia, with the support of the World Bank. It was not clear why the MoA had not assumed lead responsibility in launching and implementing this initiative when soil fertility goes beyond fertiliser use. The National Fertiliser Industry Agency has pursued the implementation of SFI in Ethiopia by introducing soil fertility management as one of the components of the National Fertiliser Sector Project, which is financed by the World Bank.

The project's dominant component is fertiliser demand, which was allocated 93 percent (US$214.22 million) of the total 12 budget (US$230 million) and the soil fertility management has 3.6 percent of the project budget. The soil fertility sub-components focused on looking at soil nutrient issues where soil testing laboratories and fertiliser trials become the main activity. Under this project, soil fertility management was pursued from a narrowly nutrient point of view in isolation from the broader socioeconomic and environmental issues.

Soil fertility management requires a broader framework to attain its objectives. Yet, the Government's National Extension Package programme with emphasis on increasing yields, has not been as flexible in responding to the various agro-ecological zones, local resource endowment and farmers' capacity to invest in affordable soil fertility management techniques. The packages are designed by research-extension experts with little or no serious effort made to integrate environmental sustainability issues (crop and plant biodiversity) as well as indigenous knowledge and practices and crop and plant diversity at community level. The package approach seems to have overplayed the production aspect through making investment on external inputs, which is out of reach of the vast number of resource-poor farmers who have no capacity to invest in this package.

Management of Forests, Woodlands and Tree Resources

Forest and biomass resources are the most indispensable source used in meeting the energy needs of the rural households in Ethiopia. They are also used for farm implements, construction and as a source of cash income. The use of forest resources by the industrial sector is not significant and industrial plantation and Government-managed National Forestry Priority areas and other protected areas are not addressed in this paper. The focus is on management of forest, woodlands and tree resources close to the farming, homestead, village and community levels as well as finding alternative biomass energy sources that is simple and affordable to farmers.

In the past Government policy did not encourage on-farm and homestead tree growing and even to this day, there is no clear policy in this regard. In some parts of Ethiopia, for example in the Tigraye Region, tree planting near and on the farm is seen as undesirable and competing with cropland. Forest research as well as agricultural extension has largely ignored on-farm tree growing. Recognising that forest utilisation far exceeds their replenishment in rural communities and that its management is closely linked to land conservation, soil fertility, water recharge, biodiversity, food security and livelihood issues, the 1997 Conservation Strategy of Ethiopia includes some basic tenets that try to address these problems.

The CSA guidelines on forest, woodlands and tree management specifically mention that: i) forest development and projects are to be executed primarily by individuals and communities and the Government's role is the provision of technical support (extension and research) and enabling policies (such as pricing policy and increased security of land and tree tenure); ii) forest development strategy should integrate the development, management and conservation of forest resources with those of land and water resources, energy resources, ecosystems and genetic resources as well as crop and livestock production; iii) efforts should be made to achieve the principle of "Sustainable Forest Management" where the volume of wood harvested is about equal to the net growth that forest is capable of generating in a socially acceptable and economically viable way.

The most crucial and immediate steps to address the depletion of forest resources that have an impact on other resources and sectors is on the supply side. Here increasing the production of forest products for fuel wood, timber, fodder and construction material is vital. The most promising vehicle to achieve this will be through individual tree planting by smallholders and agroforestry devolvement. This would also bring added benefits to mitigating land degradation (through the increased of vegetative cover) and to reducing the high cost incurred in the loss of crop productivity by the conversion of dung and crop residue for energy needs.

Community forestry, if managed in a participatory manner, has the potential to contribute meaningfully to increase forest products and meet the needs of rural households. In the 1980s the previous military Government introduced community forestry to meet the fuelwood and construction needs of rural households. Farmers for the most part were ordered to plant in the community forest (by peasant association) with no clear guidelines whether they would benefit from the trees planted and the grass growing from the community forestry. Needless to say, most farmers saw community forestry as an extension of the Government owned afforestation scheme.

In fact, in some places the Government converted community forestry into protected forest areas. The net result was that community forestry was generally poorly managed, with very low survival rate to meet the expected needs. With the fall of the Government in 1991, this approach was abandoned. Recent experience in some regions in Ethiopia has shown that collective action in managing woodlots has worked well if they were managed at the lowest administrative level, involved villagers in decisionmaking and benefits were shared fairly among participants.

Renewable energy sources, particularly in the form of biomass energy, are untapped in Ethiopia. The Conservation Strategy of Ethiopia makes very brief reference to the need for development of alternative energy sources namely, solar, wind, biogas, agricultural biofuel for small towns and villages. However, little investment has been made in this regard, particularly biomass energy as it is widely available in most rural areas. Urgent attention and priority is required to examine the feasibility of converting agricultural and crop residue, dung and other wastes into an efficient form of energy. In this regard, biogas generation through anaerobic digestion of dung into high-energy form (methane) that could be used for cooking and lighting should be given priority. The by-product from this process (the slurry) would result in a good quality and environmental friendly fertiliser that can be used to enhance soil fertility. This process also contributes to improving sanitation at the village level, as experienced in China.

Water Harvesting in Rural Areas

Until very recently Government policy had not accorded a significant role to rainwater harvesting (RWH). The recently approved Rural Development Policies recognises the importance of rainwater control and utilisation around the farmers' plot and the work to be done by the farmer himself through labour-intensive technologies. This is a positive development although there is still confusion on the definition of rainwater harvesting since small river diversion is also included as part of rainwater harvesting.

Rainwater harvesting, particularly for agricultural and natural resources management (which is the focus here) is closely linked to agronomic practices, farming system and livelihood activities of the communities. The aim is to make rainwater harvesting a collective group effort for those who live in close proximity to each other in a village or community as is the case in most parts of rural Ethiopia. Groups will also have greater economy of scale to reduce costs incurred in RWH technologies (pooling labour and resources) and reduce health risk through joint monitoring and maintenance of the system.

Many of the experiences gained in RWH (either through WFP, NGOs or MoA) have been at homestead level. During a recent mission to Ethiopia, field visits were organised by MoA to see some of the pilot activities on Homestead Rainwater Harvesting (that involves individual farming families through support from Food-for-Work Programme and MoA) outside the city of Nazareth. Most households in this dryland and marginal areas do not have adequate food to last them during the long dry season.

The benefits of RWH were quickly noticeable in terms of reduced distance and time on women and children in fetching water, improved nutritional status, ability to expand agricultural, horticultural, livestock, chicken and beekeeping activities particularly during the dry season, improved soil conservation and vegetative cover and environment around the homestead. One farmer interviewed indicated RWH had helped him to grow onions, vegetables and fruits, which he sells in Nazareth and in return buys food items that last him during the dry seasons.

The RWH system, (using roof and micro-catchments/ farmers field) has made a difference in improving the nutritional status and diversifying the livelihood of farmers. Many of the experiences gained in RWH (through WFP, NGOs or MoA) have been at homestead level. Preliminary observation and discussion with farmers and local extension suggest that individually centred RWH for domestic and non-domestic purposes (agriculture, livestock, etc.) faces serious constraints to being a viable option without considerable external financial investment and supervision.

Most farmers in this area, for example, do not have a tin roof, seriously limiting the amount of roof water to be harvested and stored. Farmers interviewed use a simple, low-cost hemispherical system (holding 50 to 60 cubic metres of water below the ground with cemented floor and walls and organic matter for roof cover) and indicated that they would not be able to cover the cost of construction material, skilled labour and transportation if it had not been for considerable outside assistance.

Extension agents also indicated that maintenance of the system is the same with the simple hemispherical tank (which has the advantage of holding more water at low cost than the other systems such as domed or the Chinese systems) and is much better if it is handled by groups of farmers rather than individual farmers. Sediments can easily enter a hemispherical tank lowering its water holding capacity and life span. It requires proper maintenance, which is difficult for individual farmers to handle and organise.

In addition, this particular system occupies a large space in farmers' fields (as it needs grass strips or trees along the water collection channels, which were not seen), taking up more farmland, which is in serious shortage in this area. Thus, this emphasis here is to see RWH as one of the essential elements/components requiring collective action to achieve a community-based natural resources management and development to enhance food security and livelihood of the rural people in a sustainable way.

Grazing Land and Livestock Management

There was no specific policy regarding livestock density or managing grazing land, until the guidelines that resulted after the Fourth Livestock Development Project in 1993-1994 and the Conservation Strategy of Ethiopia in 1997. Livestock and overgrazing have adverse impact on soil degradation, compaction and reduction on vegetative and biomass cover. It is certain that a policy limiting the number of livestock would not be popular or difficult to implement since they are the most significant means of capital accumulation and quickly disposable assets in time of famine and other emergencies. Even farmers indicating serious land shortage did not like their livestock size to be reduced.

Some of the past as well as the current Conservation Strategies of Ethiopia, such as hillside and area closure, have underestimated the role of livestock and grazing practices in the process of natural resources degradation. Severely degraded areas with no or little vegetation have no human and livestock interference for 3-5 years until 80 percent of the grasses or vegetation are regenerated. Farmers in general are negative towards hillside and area closures (particularly in communities where even cut-carry is not practised) as it has meant taking away the grazing land in their communities forcing them to use their own productive land thereby exposing it to further degradation.

The Highland Reclamation Study and its recommendations, which have greatly influenced the Conservation Strategy in Ethiopia, mainly focus on crop land. The severity and incidence of soil erosion (due to absence of vegetative cover) on cropland more than grazing land is not disputed. But the role of livestock cannot be neglected since nearly half of the country's land is used for grazing and during the major rainy season livestock graze on the slope leading to "downstream effect" where water erosion and flooding affects agricultural land. In addition, cropland is left open for grazing during the long dry season denuding all vegetative cover and causing soil crusting, which exacerbates the incidence of soil erosion and decline in soil fertility.

Past efforts in livestock improvement may have focused on promoting veterinary service, which is important. But the major factors resulting in low productivity of Ethiopian livestock sub-sector is the serious shortage of feed. Livestock depends on natural pastures that are low yielding and grazing areas are shrinking due to expansion of cultivated land. Hence, improved livestock feed and forage production is crucial from both food security and prevention of natural resources degradation.

The Fourth Livestock Development Project launched in 1988 by the MoA focused on improved forage development species that are low cost, acceptable to farmers and also contribute to soil and water conservation. These included backyard forage, under sowing forage legumes into cereals, stock exclusion area, forage strip establishment (soil and water conservation), and annual and perennial forage establishment. It has also given emphasis on the need to shift from shifting and uncontrolled and environmentally degraded grazing practices into a more intensive livestock feeding and management system.

One of the most notable Government strategies regarding livestock improvement is the National Livestock Development Programme (NLDP) which consolidates and expands (with appropriate modification) the results obtained from the Fourth Livestock Development Project, particularly in the area of forage and seed production that are suitable to local conditions. The NLDP has come up with modified forage development strategies for different agro-ecological zones and farming systems that could also be used by other development agencies involved in natural resources management activities.

7

Land Quality Indicators

The UN Commission on Sustainable Development (CSD) in its work programme indicated the need for indicators of sustainability linked to each issue discussed in Agenda 21. Among the well over 100 indicators defined, two refer to Agenda 21 Chapter 10, the Integrated Planning and Management of Land Resources. These are Land Use Change and Land Condition Change (change in the aggregate of the land qualities).

It will be clear that the land condition indicator subsumes everything that is being discussed in the present Consultation on Land Quality Indicators. Here lies the core problem of the CSD effort. Unlike the relatively narrow economic indicators traditionally used by governments, central banks and the World Bank and regional development banks, which can summarize various aspects of the economy in a single, monetary variable, the land condition indicator is essentially multi-dimensional (as is the land use indicator).

The monitoring of broad changes in land use and in general land conditions will enable governments to report on the general improvement or deterioration of the state of the natural resources in their countries, using standard methods that allow comparisons between countries and between years. However, such broad, very highly aggregated indicators will not be useful in support of policy development, either national or at more detailed scale.

For practical purposes of policy advice or management advice, much more detailed, disaggregated indicators will be needed. In fact, the more detailed and specific the information requirements, the less the concept of land quality indicators makes sense, and the more evident is the need for actual observations on specific land use changes and on specific aspects of land condition, of physical, biological and chemical soil conditions, and of plant nutrient stocks and availability of plant nutrients in soils.

Land Use Changes

Is land use change an indicator of a driving force, state or response? Land use changes are generally conscious, volitional responses by humans or human societies to changes in biophysical or societal conditions. It is a response indicator, therefore, reflecting how and to what extent society is responding to meet its changing needs and goals or to adapt to changing environmental conditions. This does not exclude the possibility that some land use changes may, in turn, constitute a driving force for changes in the state of the environment. That is in the very nature of the complex causal network (not a simple causal chain), including a number of feedback loops, that is society's relationship with its environment.

The rates or extent of land use changes can be represented in the form of a transition matrix of land uses, showing the transition between each pair of uses as extent or proportion of area per unit time. An example of such a transition matrix can be found in the Forest

Resources Assessment 1990 for different kinds of forest and woodland versus broad categories of non-forest. In the absence of a widely accepted, systematic set of land use definitions, it was not possible to compare land uses between countries or continents except in the most general terms.

The concept of land use as a sequence of management activities in relation to a given area of land has enabled FAO, jointly with several partners, to embark on a computer-aided system of land use description usable at several levels of detail, and on an effort to arrive at a system of land use classification that serves as a standard translator between different local or regional land use information sources.

A major part can be obtained by interpretation of remote sensing data with limited ground verification to produce land cover maps (land cover is the physical vegetation or crop cover resulting from natural processes or human activities). The land cover data then need to be complemented through more detailed field work with the more specific land use information (the management activities that characterize the system of use) to arrive at a land use map.

As is the case for land use change, it is doubtful whether a single, aggregate measure of land condition or land condition change would be feasible. What is feasible in principle is an estimation of the change in the different land qualities that influence the suitability of the land for one use or another, or for conservation - for example, of biodiversity.

The data needed for interpretation of changes in land qualities can be derived in part from remote sensing, complemented by ground survey observations. These should be combined with more frequent and thorough monitoring at selected permanent benchmark sites and long-term land productivity trial sites. Information is needed on aspects such as nature and density of vegetation or nature and productivity of crops (partly overlapping data with land use); conditions of the land surface (slope, runoff and erosion rates, sediment transport by wind, surface salinity); hydrological conditions (flooding incidence, groundwater dynamics); and physical, biological and chemical conditions (toxins, nutrients) of the soil.

Such monitoring information, superimposed on a baseline soils and terrain database, will allow the estimation of potential productivity changes of land under different main uses and of suitability changes for different purposes such as conservation of native vegetation or animal populations. The institutional framework that would make such systematic, repetitive monitoring feasible does not yet exist in many countries.

Soil

The soil, as a major subsystem of land, is changing with time consequent on changes in its environment (e.g., rainfall) or in management (grazing intensity, crops, irrigation, inputs, etc.). Land use or management decisions at any scale, from individual farm family to nation, need information about the soil system and its relationships with the environment and with management options. These relationships, as in the case of land use, are complex, with feedbacks providing different metastable situations and non-linear, sometimes delayed, responses to change. This implies that some changes in external conditions may not cause a perceptible change in the soil condition, while others may cause acute, or gradual, degradation or improvement.

Figure 1 shows the complex of processes that may lead to more biologically active and productive soils with a gradual rise in concentration of atmospheric carbon dioxide, as has started in the last century and is expected to continue for at least several decades. Clearly, a simple soil condition indicator cannot capture such processes. That will require both modelling and monitoring of a range of variables.

Even in a less complex system, in which monitoring shows increasing runoff and consequent gully formation in sloping soils, for example, such a change in soil condition cannot by itself be used to diagnose the proximate or underlying causes and does not contribute to identification of the most appropriate responses. These might be as diverse as advising the land users to change crop rotations or plant dense contour lines of permanent crops, or improving access roads or markets in the area, or enabling the users to obtain secure, long-term land tenure. In every case, the system as a whole will need to be understood rather than one possibly diagnostic or indicative element, before sound responses can be identified and put into practice.

Plant Nutrients

Descending another step, into the plant nutrients or soil fertility subsystem of soil, one finds similar complexity. Nitrogen deficiency, for example, may be caused by a low content of decomposable organic matter in the soil, or by a high groundwater table, and remedied by applications of manure or appropriate fertilizer or green manuring, or by land drainage. Similarly, zinc deficiency in wetland rice may be caused by an absolute zinc deficiency, or by poor availability of zinc due to fixation in the presence of calcium carbonate, or by very gradual, but strong fixation

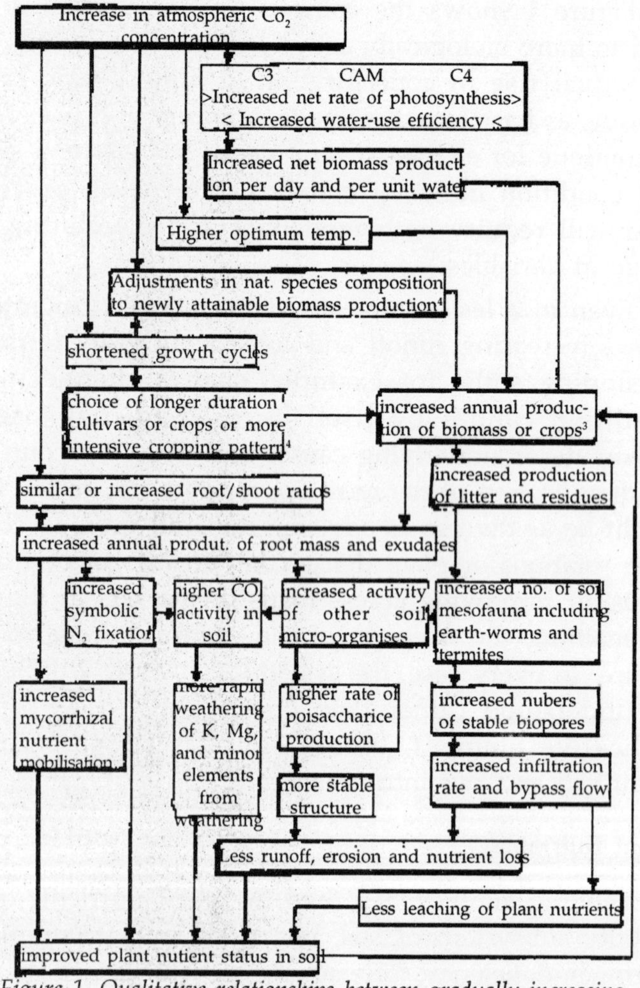

Figure 1. Qualitative relationships between gradually increasing atmospheric CO_2 concentration, soil characteristics[1] and medium-term processes in soils[2], and biomass or crop productivity[3]

[1] Gradually rising CO_2 as in this century and in transient global change scenarios. [2] Soils with some weatherable minerals at least in the subsoil or substratum within rooting depth. [3] Extreme weather events may disrupt some relationships in the figure, so any major increase in their frequency or intensity may counteract positive effects shown. [4] Species composition adjusts, or choices indicated are made to adjust, to the newly attainable biomass or crop production under increased atmospheric CO_2, compensating for shortened growth cycles of existing species or crops.

where soils remain continuously wet and reduced for long periods, without drying between rice crops. So even at this level of detail, single indicators may be meaningless until the related factors in the system are known and the system as a whole is clearly understood.

8

Land Degradation

Soils will naturally evolve and change over time as a result of such pedogenetic processes as leaching and erosion. In areas of undisturbed natural vegetation, such changes are generally slow and in a historical time frame will have little effect on the ability of the soil to support vegetative growth. However the natural processes may be accelerated in areas exploited by man for agricultural purposes, thereby producing, often within only a few years, major changes in the soil biological, chemical and physical properties.

Such a change in soil properties will alter, and usually reduce, the land's ability to sustain a particular quantity and quality of plant growth. From an agricultural point of view any such reduction can be regarded as degradation. The ability of land to support specific forms of agricultural production is finite. Important limits to production are set not only by the soil properties but also by such factors as climate, relief, hydrology, vegetation and land use. Land mismanagement—whether for crop, livestock or tree production purposes—consists usually of removing too much, returning too little and cultivating, grazing or cutting too often.

Such "mining" of land beyond its limits results in degradation with decreasing productivity, and is non sustainable. For any given land area there are limits on the types of land use that can be pursued on a sustainable basis. Likewise their potential production will be limited according to the management practices and input levels.

Soil and Land Degradation

In the soil conservation arena the terms soil degradation and land degradation are sometimes incorrectly used interchangeably, with soil erosion regarded as synonymous to both. However there is more to soil degradation than just soil erosion, and land represents a broader concept than simply soil. As with its use in the context of land evaluation, the term land refers to all natural resources which contribute to agricultural production, including livestock production and forestry. Land thus covers climate, landforms, water resources, soils and vegetation (including both grassland and forests).

Whereas combating soil degradation is critical to SARM, this cannot be addressed in isolation of other natural resources, as degradation of one can be expected to have an adverse impact on the agricultural productive capacity of the others. Following the 1992 UN Conference on Environment and Development (UNCED), desertification has been defined as "degradation of land resources in arid, semi-arid and dry sub-humid areas, caused by different factors including climatic variations and human activities resulting from adverse human impact". It needs to be stated that desertification is not a separate form of land degradation, so much as the end product of a variety of degradation processes that have adversely affected the land within so-called "dry" areas.

Desertification is an emotive term that conjures an image of desert sand dunes advancing over adjacent areas, something that rarely happens in reality. Also, what at first sight might appear to be desertification (e.g. loss of vegetative ground cover) may be a cyclical feature associated with a sequence of drought, or lower-than-average rainfall years. Vegetation recovery can be dramatic once relatively wetter rainfall conditions return. Placing undue emphasis on the threat of desertification can lead to the erroneous assumption that desertification and degradation are one and the same thing. This ignores the fact that land degradation can also occur under humid climatic conditions where the economic impact in terms of lost production greatly exceeds that of drier areas.

Three factors are commonly cited as causes of desertification: overgrazing, inappropriate agricultural practices and overuse of woody biomass. Such factors are not confined to dry areas and can lead to equally severe degradation in humid areas. Land degradation following mismanagement of the natural resources can occur anywhere, irrespective of the prevailing climatic conditions. For this reason it is believed that the emphasis placed on desertification programmes within the Asia Pacific region is misdirected. What is needed is a commitment to a holistic approach to land degradation wherever it might occur.

Land degradation is the loss of the productive capacity of the land to sustain life. However such a definition is perhaps too broad and has somewhat emotive overtones. It ignores the fact that whereas the productive capacity of an area may have been reduced by land degradation, it may still be possible to use the land for productive purposes by adopting alternative land uses, although with an inherent lower productive potential. While land degradation will have taken place, it will not have progressed to the extent that the land can no longer sustain any form of life. It therefore may be more appropriate to define land degradation in a more focused manner as follows:

Components of Land Degradation

There are a number of interrelated land degradation components, all of which may contribute to a decline in agricultural production. The most important are:

— *Soil degradation*—decline in the productive capacity of the soil as a result of soil erosion and changes in the hydrological, biological, chemical and physical properties of the soil.

— *Vegetation degradation*—decline in the quantity and/or quality of the natural biomass and decrease in the vegetative ground cover.

— *Water degradation*—decline in the quantity and/or quality of both surface and ground water resources.

— *Climate deterioration*—changes in the micro and macro climatic conditions that increase the risk of crop failure.

— *Losses to urban/industrial development*—decline in the total area of land used, or with potential, for agricultural production as a result of arable land being converted to urban, industrial and infrastructure uses.

A joint FAO, UNEP and UNESCO study defined soil degradation as a process which lowers the current and/or the potential capability of the soil to produce (quantitatively and/or qualitatively) goods or services.

The study regarded soil degradation as not necessarily continuous but something that could take place over a relatively short period between two states of equilibrium. For instance clearing an area of forest and then using the land continuously for low input maize monoculture result in a rapid decline in the soils humus and nutrient levels.

Providing soil erosion does not physically destroy the resource, soil cultivated in this way would not attain or even closely approach a zero humus and nutrient content. Instead it would reach a low-level equilibrium in which humus and nutrient levels remain constant and crop yields are stabilised at a low level. Many of the soils used for low input agricultural production in the Asia Pacific region are believed to have reached this low-level equilibrium.

The FAO, UNEP and UNESCO study recognised six categories of soil degradation processes:

1. Water erosion
2. Wind erosion
3. Waterlogging and excess of salts
4. Chemical degradation
5. Physical degradation
6. Biological degradation

Water Erosion

Water erosion is the most widespread form of degradation within the Asia Pacific region and occurs widely in all agro-climatic zones. This category includes processes such as splash erosion, sheet erosion, rill and gully erosion and mass movement.

Splash erosion commonly initiates water erosion and occurs when raindrops fall onto the bare soil surface. The impact of raindrops breaks up the surface soil aggregates of Papua New Guinea, landslides may be natural phenomena. However their frequency and severity may greatly increase following destruction of the natural vegetative cover by logging and/or clearing for cultivation, as in the Sierra Madre range in the Philippines.

Wind Erosion

The risk of wind erosion is severe in the arid and semi-arid areas of Asia and Australia. It includes both the removal and deposition of soil particles by wind action and the abrasive effects of moving particles as they are transported. It occurs when soil is left bare of vegetation as a result of cultivation, and/or overgrazing following overstocking. Not only can the wind remove topsoil from good farmland; it can result in additional damage by burying land, buildings, machinery and fences with unwanted soil. It is estimated that in Pakistan some 42% of the arable land are affected by wind erosion. In India the figure is 6%, although the total area affected, 11 million ha, is the same as for Pakistan. In China there are reports that windblown sand has affected some 2.65 million ha of cultivated land.

Waterlogging and Salinisation (excess of salts)

Waterlogging is the lowering in land productivity through the rise in groundwater close to the soil surface. In its most severe form (termed ponding), the water table rises above the surface. Within the Asia Pacific region the most severely affected countries are India and Pakistan where some 3.08 and 2 million has respectively have been affected. Waterlogging is linked with salinisation, both brought about by incorrect irrigation management.

Salinisation is used in its broad sense, to refer to all types of soil degradation brought about by the increase of salts in the soil. It thus covers both salinisation in its strict sense, the build-up of free salts; and sodification (also called alkalisation), the development of dominance of the exchange complex by sodium. If topsoil becomes too saline or too alkaline its productivity falls.

The processes of salinisation can happen when poorly drained land is irrigated in hot climates. The sun evaporates the surface water, leaving behind the salts. At the same time, inadequate drainage causes the water table to rise, bringing saline groundwater into contact with plant roots. It is typically a human-induced process brought about by incorrect planning and management of irrigation schemes.

Saline intrusion is a localised form of salinisation which occurs as the result of the incursion of sea water into coastal soils arising from over-abstraction of groundwater. This is of particular concern to several Pacific nations in the context of rising sea levels. One very serious effect of such a sea level rise is its impact on freshwater lenses underlying atolls. The risk of saltwater intrusion will rise as the sea level rises; lateral leakage will increase, the lenses will become thinner and salt water will move within reach of pump intakes. As the sea level rises salt water will reach the roots of pit grown taro, coconut palms and other tree crops.

Chemical Degradation

In addition to salinisation other processes may adversely affect the chemical properties of soil, notably:

- Loss of nutrients and organic matter;
- Acidification (mostly associated with the removal of soil nutrients or the misuse of fertiliser);
- Fixing of soluble nutrients, so that they are 'unavailable' to plants, following acidification;

— Increased leaching (when the vegetative cover is removed and the soil exposed particularly when this occurs in high rainfall tropical areas);

— Increased temperatures and oxidation of the soil organic matter (due to exposure and cultivation); and

— Pollution (most commonly from improper management of industrial and mining wastes.

Of particular concern for SARM is the reduction in the reserves of soil nutrients. When soils are used for agricultural purposes significant quantities of nutrients are removed from the harvested products. With the present level of world yields the annual amount of harvested plant nutrients in three major cereals (rice, wheat, maize) is estimated at 40 million tons of N, 15 million tons of P2O5, and 28 million tons of K2O. If nutrients removed are not replaced (in the form of chemical fertilisers, organic manure, by natural fixation from the air or by weathering of rock minerals), then there will be a net decline in soil nutrient levels.

Plant growth may be adversely affected by the build-up of particular metals or salts to toxic levels within the soil. Aluminium toxicities may be a problem in strongly acid soils. The higher aluminium levels found in the subsoils of some tropical soils may be an additional factor in yield decline following loss of topsoil from erosion. Calcium carbonate and gypsum (when present as calcrete or gypsum horizons) may cause nutrient imbalance. Manganese toxicity is most likely in acid soils with impeded drainage.

Several countries in Asia have a significant proportion of their land area with acid sulphate soil limitations, notably Cambodia (1.2%), Malaysia (2%), Thailand (2%) and Vietnam (4.6%). Such soils are commonly developed on estuarine and marine alluvium (mangrove swamps) and contain considerable amounts of sulphides. Chemical degradation can be rapid once such areas are "reclaimed." Once drained, the sulphides oxidise to form sulphuric acid; and the pH, which is around neutral prior to drainage, drops below 3.5. The acid attacks the clay minerals causing the liberation of aluminium ions in amounts toxic for plant roots and micro-organisms.

Toxicities can also result from the presence of municipal, industrial, radioactive or oily wastes such as occur mainly around towns, industrial areas and mines. Such toxicities are usually of only local extent but in some countries they may lead to significant problems, for instance the extensive areas of tin tailings from former tin mining operations in Malaysia. In Vietnam (and possibly Laos and Cambodia) there are areas where the soils contain residues of the chemical defoliants used between the 1960s and 1975, during the American phase of the Indochina war.

Physical Degradation

Both crop and livestock production can lead to a deterioration in the physical condition of the soil. This degradation can take many forms, and has a variety of consequences. Deterioration of soil structure is the most common form of physical degradation, and involves:

- — Loss of stability of aggregates in surface soils, leading to crusting and compaction, and consequently poorer infiltration rates and greater run-off and erosion; and
- — Translocation of clay particles to subsurface layers and loss of porosity in the subsurface and deeper soil layers, with consequent loss of water transmission and storage capacity.

Physical degradation is of concern because soil structure and its stability governs soil-water relationships, aeration, crusting, infiltration, permeability, runoff, interflow, root penetration, leaching losses of plant nutrients and therefore ultimately the productive potential of a soil.

Topsoil degradation occurs when an open structure of soil aggregates is broken down by excessive tillage. Also, the impact of raindrops and/or livestock hooves may produce a continuous compacted layer or crust at the surface. Reduction in topsoil porosity, and particularly surface crusting, will result in decreased water infiltration, increased runoff, poorer seedling emergence and often increased erosion.

Physical degradation of the subsoil may be in the form of a distinct pan, or a more general compaction, i.e. loss of the original subsoil structure and increasing bulk density with reduction in size and quantity of pore spaces. The physical effects may be decreased water storage capacity, loss of aeration and reduced soil permeability. Waterlogging may occur in the soil above the compacted horizon and the absorptive capacity of the subsoil will be reduced, thereby increasing the amount of rainfall going as surface runoff. Plant root development will be hindered in the subsoil because compacted horizons are physically difficult to penetrate. Plant growth will be restricted because of the lower availability of air and water, and therefore nutrients, in the root zone.

Some of the blame for increased soil degradation in the region has been ascribed to a specific change in farming practice, namely the replacement of the traditional digging sticks and hoes with the plough. In part, this has facilitated the expansion of the area under cultivation thereby exposing more land to the risk of accelerated erosion. The main reason is that ploughing, using techniques imported from temperate regions in Europe and America, increases the disturbance of the topsoil, breaking up its structure and making it less resistant to erosion.

Traditional shifting cultivation systems rely on the presence of tree stumps and roots, not only for rapid restoration of the vegetation during the fallow period, but also for reducing runoff velocity and holding the soil in situ during cultivation. The conservation benefits of the traditional land clearing techniques are lost once the plough has been adopted, as ploughing is difficult, if not impossible, unless the stumps and roots are first removed from the field.

Subsoil compaction, particularly the formation of "plough pans," is normally associated with commercial agriculture and the use of heavy machinery such as tractors. However, evidence suggests that a cultivation pan, immediately below the cultivated layer, can occur even under traditional hand cultivation systems from the pressure of the tiller's feet and the practice of cultivating to the same depth each year (Shaxson personal communication). Although the low yields under low input systems may largely be because soil nutrient and organic matter levels have reached low-level equilibrium, physical degradation in the form of a cultivation pan may also be a contributory factor.

Whereas conservation measures can be adopted to control soil erosion, and chemical fertilisers can be used to replace soil nutrients, physical degradation, particularly in the subsoil, is less easily overcome. Subsoil compaction is regarded as a form of soil degradation as it markedly reduces the yields of dryland crops and may be expensive to overcome, requiring deep ripping with a tractor. However in areas where paddy rice is grown, farmers adopt soil compaction techniques, such as puddling, to reduce subsoil permeability thereby raising the capacity of the soil to produce rice. In such a situation soil compaction is regarded as land improvement rather than degradation.

Biological Degradation

Soils used for agricultural purposes are often deficient in the biological processes which both maintain their physical structure and their ability to supply essential chemical elements to plants. Of particular concern is the decline in soil organic matter or humus content following cultivation. In part this is because large amounts of biomass are harvested and removed from the site. The actual humus mineralisation rates may increase due to soil temperature changes following the removal of a protective vegetative cover.

The agricultural significance of soil organic matter is greater than that of any other property with the exception of soil moisture. It improves soil structure, and thereby root penetration and erosion resistance; augments cation exchange capacity; and acts as a store of nutrients, slowly converted to forms available to plants. It is possible to obtain an overall balance of soil organic matter with shifting cultivation under conditions of low population density.

However for most of the Asia Pacific region shortage of suitable land means that shifting cultivation is no longer a viable option and most of the land is cultivated each year. Under such permanent agricultural systems decline in organic matter can be severe and rapid. Typical values of the organic matter. status of tropical soils that have been under cultivation for two or more years are 30-60% of the corresponding values under natural vegetation. Values below 50% are considered to represent an undesirable situation calling for remedial measures.

Biological degradation is usually synonymous with decline in organic matter. Yet it also applies when there is a decline in the beneficial soil fauna (organisms). Certain soil organisms have the ability to influence the physical structure of soils. Some species of termites annually transport large amounts of soil through the soil profile. In so doing they achieve a mixing of the organic and mineral components of soil and alter the porosity of the soil at the micro- and macropore levels. This can result in increased surface infiltration of rainwater and a consequent reduction in runoff and erosion, which (together with the stimulation to soil aeration) contribute to maintaining soil fertility. Earthworms, while important for the soils of the temperate parts of the Asia Pacific region, may play a similar role as termites in some soils in the tropics, although not in comparable numbers or biomass.

Most mechanical forms of soil tillage bring about marked changes in the qualitative and quantitative characters of the soil fauna community. The consequences of such biological degradation are poorly understood and merit further study. However the improved soil structure and enhanced efficiency of nutrient return from crop residues to plant, associated with conservation (minimum) tillage techniques are believed to be associated with the maintenance of a healthy soil fauna, particularly of termites and earthworms.

Soil Fertility

Whereas the deterioration in the chemical, physical and biological properties of the soil is discussed separately in the previous three sections their combined effects can be considered under the heading of soil fertility decline. Decline in fertility is a major effect of erosion and the term is best used to describe the combined effect of processes other than erosion. The main processes involved are:

— Lowering of soil organic matter, with associated decline in soil biological activity;

— Degradation of soil physical properties (structure, aeration, water holding capacity), as brought about by reduced organic matter;

— Adverse changes in soil nutrient resources, including reduction in availability of the major nutrients (nitrogen, phosphorus, potassium), onset of micronutrient deficiencies, and development of nutrient imbalances; and

— Build-up of toxicities, primarily acidification through incorrect fertiliser use.

Such a decline in fertility has occurred in some of Tonga's rich volcanic soils, where the large-scale expansion of squash cultivation has led to unprecedented tree removal and the indiscriminate use of fertilisers and pesticides. It has been reported that an increasing number of growers are finding that no matter how much fertiliser or pesticide they apply their yields are dropping. The damaged areas are referred to as "Hot spots ... big areas of land that have been cleared with hardly any trees left, and where the land has been farmed continuously for a number of years ... so that the structure of the soil in these areas has been destroyed and the soil no longer can absorb water to feed the plants".

A study of land degradation in South Asia noted that an interrelated set of soil fertility problems had been reported, directly or indirectly associated with fertiliser application.

— *Organic matter depletion*—crop residues are widely used as fuel and fodder and not returned to the soil. Response to fertiliser may be lower as a consequence of decreased organic matter.

— *Continuing negative soil nutrient balance*—removal of nutrients from the soil in crop harvest appears substantially to exceed inputs as natural replacement and fertilisers. Negative soil nutrient balances have been reported for all three major nutrients in Bangladesh and Nepal; for phosphorus and potassium in Sri Lanka, and a large deficit for potassium in Pakistan. For India it has been estimated that the nutrient deficit is 60 kg/ha per year, or 9 million tons for the whole country.

— *Imbalance in fertiliser application*—Fertiliser use is dominated by nitrogen. When fertilisers are first applied to a soil, a high response is frequently obtained from nitrogen. The improved crop growth depletes the soil of other nutrients; "In such systems, nitrogen is simply used as a shovel to mine the soil of other nutrients".

— *Secondary and micronutrient deficiencies*—Sulphur deficiencies have been reported in Bangladesh, India, Pakistan and Sri Lanka and zinc deficiencies in Bangladesh, India and Pakistan. Micronutrient deficiencies are being increasingly reported in Pakistan.

— *Failure of increases in fertiliser use to be matched by increases in crop yield*—A levelling off, or plateau, in the crop yield increases which took place in the 1960s and 1970s has been found in many countries of South Asia.

— *Lower responses to fertilisers*—long term experimental work in India has shown low or zero response to particular fertilisers when one or more other nutrients have been deficient.

Although the study focused specifically on countries in South Asia similar problems associated with the Green Revolution emphasis on the use of chemical fertilisers for increasing crop production can be expected to occur in most if not all the Asia Pacific countries.

A particular problem associated with the increased use of nitrogenous fertilisers is that, irrespective of how efficiently it is applied, nitrogen recovery efficiencies greater than 50% are rarely achieved. Much of the unrecovered nitrogen ends up in groundwater in the form of nitrate,13 or in the wetlands as ammonia in the atmosphere. When it is washed off the soil surface (or through extremely sandy soils) into waterways it may cause eutrophication problems, proliferation of algal growth and exhaustion of dissolved oxygen with consequences for fish and other aquatic species.

Vegetation Degradation

Vegetation degradation is usually regarded as a reduction in the available biomass, and decline in the vegetative ground cover, as a result of deforestation and overgrazing. Such degradation is a major contributory factor to soil degradation particularly with regard to soil erosion and loss of soil organic matter. The term also applies in situations where the reduction is not in the quantity of biomass but in quality—for instance bush encroachment into rangelands, and the loss of palatable pasture grasses and their replacement with nonpalatable species.

In such a situation the value of the land will have declined from an agricultural point of view with a decline in its livestock carrying capacity. However the degraded vegetation may still be contributing to the soil in terms of ground cover and organic matter. In practice soil degradation usually accompanies rangeland degradation but it may not be such a clearcut relationship as that associated with deforestation.

A particular form of vegetation degradation affecting significant areas within some Asia Pacific countries, notably Indonesia, Papua New Guinea, the Philippines and Vietnam, is the replacement of tropical forest with Imperata cylindrica grasslands. This typically follows forest clearing for agriculture. Poor soil management practices then lead to the onset of soil degradation with the Imperata invasion of the farmland. It is an extremely difficult weed to control by hand labour alone and has an adverse effect on crop yields often resulting in the abandonment of the field.

Deforestation

Natural ecosystems are exceptionally important for island societies, even though their total area may be minuscule. Small island forests, for example, can be absolutely critical for life support systems, especially for water supplies, subsistence wood, food and medicine, and soil stabilisation. The watersheds of small islands are far smaller than those of the Asian continent and smaller than those on the larger islands in the Asian Archipelago nations. On these smaller islands (in both Asia and the Pacific) even slight forest degradation can destroy watershed functions.

Deforestation as a result of timber extraction and the establishment of plantation crops has led to significant anthropogenic environmental transformation within the small islands of many Asian and Pacific nations. The typical process has included:

— Logging of valuable hardwoods.

— Replacement of natural forests by erosive and pest/disease-prone plantation agriculture.

— Soil exhaustion and plantation epidemics.

— Market or price collapse for plantation products.

— Plantation collapse and the subsequent marginalisation of poor people to upland forests.

— Upland deforestation.

— Consequent upland erosion, and hence further appropriation of the island's natural ecosystems—and subsequent further degradation.

— Economic disinvestment of the degraded island interior in favour of coastal development or, where environmental degradation leads to collapse of life-support systems, emigration.

— Extreme lack of investment in managing the ever-diminishing forests. And

— Dependence on remittances from migrant labour, aid and foreign investment.

— In the Pacific the process has often been exacerbated by the following:

— Natural disasters such as typhoons which periodically destroy the vulnerable plantation monocultures and remaining forest;

— Neglect of 'unprofitable' islands on the part of the central government and donors; and

— Inability of small island populations to muster the skills and the political and economic power to counteract the trend.

— As a result, biological productivity, diversity and resilience have diminished, and much land in many islands now lies derelict.

Agrodeforestation

An environmental report on the Pacific described agrodeforestation as a major threat to sustainable development. The term deforestation is well known in the context of the destruction or removal of forests. The term agrodeforestation covers the loss of trees from within existing agricultural landscapes, an issue that has received far less attention. Trees are an important component of many Asian and Pacific farming systems. With a combination of both cultivated and wild species there is considerable bio-diversity to be found amongst the tree species within individual traditional farming systems. The different tree species may be:

— Widely cultivated or deliberately planted within farm fields, on boundaries or around the homestead;

— Protected or preserved when clearing garden areas; and/or

— Protected upon spontaneous generation within the farm.

In the Pacific on the larger Melanesian islands over 100 tree species are commonly integral parts of rural and urban agricultural systems, whereas atoll systems have up to 30-40 different species. Such trees serve at least 12 distinct ecological functions, have over 70 cultural and economic uses, and provide up to 70% of the real income and production of rural Pacific people. Replacing the products from these trees with imported substitutes would either be impossible or too expensive. Elimination of these trees thus constitutes a major ecological, cultural and economic disaster which would undermine self-reliance and sustainability in the Pacific islands.

Agrodeforestation is a problem in Asia and the Pacific as agriculture becomes more intensive and specialised.

When cultivation is undertaken by hand, retaining or planting trees within fields is not a problem, but once tractors are used, then trees hinder tillage operations and are usually removed. Within a traditional mixed biodiverse farming system the number of trees per unit area may be high, whereas the numbers of an individual species may be low, given that a farm family may be able to meet its subsistence needs for a particular product from just one or two plants.

As farm families become more commercially oriented there is a tendency to specialise in the production of a more limited number of cash crops (annuals and/or perennials). This results in the removal of many existing trees as they are perceived as occupying land that could be used for more "valuable" purposes. If trees are planted then they tend to be in monocultural plantations (coffee, cocoa, rubber, coconut etc) with a resulting loss in biodiversity and the ecological benefits of the traditional mixed home garden/farm forest.

Examples of agrodeforestation can be drawn from Fiji, where trees of cultural and ecological value that were traditionally almost always protected when clearing fallow land for new gardens are now disappearing from the agricultural landscape.

Water Degradation

Land degradation, particularly soil and vegetation degradation, has resulted in a deterioration in the quantity and quality of both surface and groundwater resources over much of the Asia Pacific region. With less vegetative cover to protect against the impact of raindrops causing surface sealing, a decline in pore spaces resulting from loss of organic matter and loss of structural stability following cultivation, less rain infiltrates the soil. Runoff increases, stream flows fluctuate more than before (in particular stream flow storm hydrographs are likely to have sharper and greater peaks), flooding becomes more frequent and extensive. Groundwater recharge decreases, streams and springs may cease and the water table is likely to drop so that wells and boreholes may dry up.

Increased runoff encourages upland erosion while an increase in severity of flash flooding encourages stream bank erosion. As a result sediment loads in rivers are increasing. The storage capacity of dams and weirs is reduced by siltation, lowland irrigation schemes are affected by silted up canals and sediment deposited in the fields, hydro-electric schemes are damaged, navigable waterways are blocked and water quality is deteriorating. High silt loads reduce the fish catch not only in inland waters but also where silt-laden rivers discharge into coastal waters.

In Thailand measurement of the suspended sediment and bedload transported by the Ping, Wang, Yom and Nan rivers suggests that each year some 27.5 million tons of soil are removed from a total catchment area of about 70,000 km^2. The Harbour Authority of Thailand reportedly dredges over 19 million m3 of sediment every year out of the last 18 km of the Chao Phya River at a total cost of 424 million baht in order to keep the channel open and to control floods.

The implications of water degradation for sustainable agriculture are serious. With less water entering the soil and stored for use during dry periods, crop yields are falling. In the drier agroclimatic zones of the Asia Pacific region this may mean the difference between success or failure in producing a worthwhile crop. In Asia a significant amount of the staple food crop production (rice and wheat) depends on irrigation, with water being supplied from dams, run of the river offtakes and boreholes. In these areas widespread catchment degradation has affected both surface and groundwater resources.

With less water available for irrigation, less land can be irrigated and less water used in individual fields with the consequence that crop yields and total production is declining. Many farm households are unable to meet their subsistence consumption needs from the production of their lowland irrigated plots. Frequently they grow a mixture of dryland crops, and/or gather forest products for sale in adjacent upland areas, to supplement their lowland farm production. The result is further degradation of the upper catchment areas contributing to yet further decline in the quantity and quality of the water resource.

A recent environmental study has identified groundwater depletion as the most ominous component of the land degradation process in Baluchistan Province, Pakistan. This is because the vast majority of both the rural and urban population of the province depend on groundwater, for which no economic alternative is available. Current levels of groundwater pumping are unsustainable with the level of the watertable dropping in some places by as much as 3 metres per year. New tubewells are continuing to be sunk leading to increased exploitation of a declining resource. If this present situation continues the study concludes that in the next 10-20 years loss of the groundwater will lead to a decline of the province's largest economic sector, irrigated agriculture, as well as the collapse of the Quetta City water supply.

In some areas the rising use of fertilisers and pesticides has led to a deterioration in water quality due to increasing contamination from a range of agrochemicals. This, combined with bacterial and chemical contamination from industrial, urban and rural (humans and livestock) sources has affected drinking water supplies. Ill health arising from water borne diseases will seriously affect the agricultural productivity of farm households.

Climate Deterioration

Although the short-term effects of land degradation are serious, evidence suggests that loss of vegetative cover and soil degradation may also be disrupting long-term rainfall patterns and increasing the likelihood of drought. For instance, the climate in the agricultural upland areas of the Northern region of Thailand appears to have become much drier following the clearing of the mixed deciduous dipterocarp forest for cultivation. Given the climatic fluctuations that occur naturally this has been difficult to prove from meteorological records.

Computer models suggest three ways in which deforestation and soil degradation may reduce rainfall:

- — Firstly, overcultivation, overgrazing and deforestation can all strip soil of vegetation. Bare soil and rock reflect more solar radiation back into the atmosphere than do grass, shrubs and trees. Increased reflectivity (albedo) keeps the atmosphere warmer, disperses cloud and reduces rain.
- — Secondly, a general lowering of soil moisture could itself suppress rainfall. Much of the rain in tropical moist forests comes from water evaporated off the vegetation, and not from outside. Wholesale clearing of rain forest breaks this hydrological cycle and may well produce a drier local climate.
- — Thirdly, deforestation and loss of topsoil structure allows the wind to throw more dust into the air.

> This dust reduces the amount of sunshine reaching the earth surface, which would have the same rain-reducing effect as bouncing more solar radiation back off the earth's surface.

Although none of the models conclusively proves the link, it is necessary to consider the possibility that (under the stresses imposed by growing population) land degradation and climatic deterioration reinforce each other.

It is theoretically possible that SARM programmes, which tackle the problems of deforestation and soil degradation, could have a positive impact on the local climate. Irrespective of possible macro climatic changes, considerable scope exists for ameliorating microclimatic conditions to improve soil conditions for the benefit of crop production. For instance the use of windbreaks, shade trees, and mulch will reduce soil surface temperatures and conserve moisture.

It has been suggested that there is a link between soil degradation and the threat of global warming from increasing atmospheric concentrations of CO_2. Terrestrial ecosystems play an important role in the global carbon budget. In addition to the effects of deforestation, world soils also have a significant impact on the global carbon budget. Rapid depletion of soil organic matter through the use of non-sustainable land use practises, may lead to emission of greenhouse gases into the atmosphere.

One estimate suggests that reduction of about 1% in organic carbon content of the top 15 cm layer of soils of the tropics could lead to an annual emission of about 128 billion tons of carbon into the atmosphere. Others question the evidence for global warming, and its possible effects on food production, believing that much of it is speculative and based on massive extrapolation and doubtful assumptions.

Should global warming prove a reality, then the expected result would be a rise in sea level and increased storm (typhoon) frequency. Some reports indicate that this is already happening in the Pacific. A 30cm sea rise is possible by the year 2030 as is a rise of 1.5°C in the temperature of the sea surface. This could result in an increase in the frequency of typhoons and their wind strength, and an increase in wave energy and destructive power, with the following physical effects:

— inundation of coastal low terrain;

— increased beach and cliff erosion;

— migration and/or reduction of wetlands;

— saltwater penetration; and

— altered tidal currents.

The lowlying island nations in the Asia Pacific region, notably Maldives, Kiribati, Marshall Islands, Tokelau and Tuvalu could be almost totally inundated. The coastal lands of many other Asia and Pacific countries are highly productive (for crop and aquaculture production) as well as the location of important infrastructure (ports, airports, roads, urban and industrial areas) and are highly vulnerable to rising sea levels and storm surges. A consequence of the threat of global warming may be the increase in the comparative advantage of areas which are less sensitive to climate change, and particularly to sea level rise. This will affect land prices and alter land use patterns, reversing the current trend towards investment and settlement in the coastal zones. Population groups can also be expected to shift to 'safer' upland areas increasing pressure on areas where the risk of land degradation is higher.

Economic development within the Asia Pacific region has led to expansion of urban and industrial land. This has been particularly rapid over the last few years in many of the newly emergent countries of Asia. Much of the present urban and industrial development has taken place on what was formerly good agricultural land. The expansion of such cities as Bangkok, Jakarta and Metro Manila has resulted in the loss of considerable areas of good quality paddy rice land.

Given the present shortage of such land, in Thailand, Java and the Philippines respectively, these losses increase the pressure on the remaining areas. Being so mountainous, with only 10% being arable land, and a population in excess of 30 million Fujian province has possibly the most acute land shortage problem in China. The uncontrolled urban and industrial development in the coastal zone (following the new policy of economic liberalisation) has seen the loss of much prime agricultural land to new roads and buildings.14

The expansion of urban settlements is a particularly acute problem in the small Pacific countries with little arable land. Population pressure is most acute in the capitals of the atoll nations (e.g. Tarawa in Kiribati) but is of concern even in those countries with larger land areas (e.g. Port Moresby in Papua New Guinea). Throughout the Pacific, people are gravitating from the mountains to the coastal cities, from the outer islands to the provincial or national seats of government, and from scattered rural hamlets to larger villages and towns. Under customary ownership laws, land is not readily available for housing estates. Hence squatter towns develop with health problems from overcrowding, unsanitary conditions and water pollution; valuable agricultural resources are lost; and forests, lagoons and reefs are degraded.

Farm households affected by urban expansion may be forced to use their remaining plots more intensively or to seek land elsewhere, which in a land scarcity situation usually means moving into marginal upland areas. Hence urban and industrial expansion may be a contributory factor to soil degradation elsewhere.

Causes of Land degradation

The causes of land degradation can be divided into natural hazards, direct causes and underlying causes. Natural hazards relate to those factors of the bio-physical environment that increase the risk of land degradation taking place e.g. steep slopes are a hazard for water erosion. Direct causes are unsuitable land use and inappropriate land management practices. Underlying causes are the reasons why inappropriate types of land use and management are practised and usually relate to the socio-economic circumstances of the land users and/ or the social, cultural, economic and policy environment in which they operate.

Natural Hazards

The major natural hazards in the Asia Pacific region, bio-physical conditions which act as predisposing factors for land degradation are:

For water erosion:

— monsoonal rains of high intensity;

— steep slopes of the mountain and hill lands;

— soils with low resistance to water erosion (e.g. silty soils, vertisols and topsoils low in organic matter).

For wind erosion:

— semi-arid to arid climates;

— high rainfall variability, with liability to drought spells;

— soils with low resistance to wind erosion (e.g. sandy soils);

— an open and often sparse cover of natural vegetation.

For soil fertility decline:

— strong leaching in humid climates;

— soils which are strongly acid, and/or with low natural fertility;

— rapid decay and mineralisation of soil organic matter in tropical climates.

For waterlogging:

— alluvial plains or interior basins which restrict outward drainage of groundwater.

For lowering of the water table:

— semi-arid to arid climates with low rates of groundwater recharge.

For salinisation:

— semi-arid to arid climates with low leaching intensity;

— plains and interior basins which restrict outward drainage of groundwater;

— soils which are naturally slightly saline;

— lowlying atolls with thin lenses of freshwater.

In some cases these natural hazards are of sufficient intensity to give rise to unproductive land without human interference. Examples are the naturally saline soils which occur in some interior basins of dry regions, or areas of natural gullying (badlands). With respect to land degradation, the key feature is that land shortage within the Asia Pacific region has led to the widespread use for agricultural purposes of areas with natural hazards. These are the passive, or predisposing, conditions for land degradation.

Erosion is a natural process. The conventional wisdom has been that soil erosion, following the growing of crops and/or grazing of livestock in upland and highland areas, is the primary cause of high river sediment levels. However, a considerable proportion of the eroded sediment found in river systems can be attributed to natural causes such as mass wasting (e.g. landslides) and various on-going geomorphological processes associated with the shaping of upland landscapes. Hence when looking for the cause of land degradation a key question that has to be asked is, what proportion of the present erosion and river sediment levels is attributable to on-going natural processes, and what proportion is largely the result of 'accelerated erosion' because of inappropriate land use?

High annual rainfall totals are a feature of many highland areas, much of which may fall within a limited portion of the year (the rainy season) and often as isolated heavy storm events. Even with excellent forest cover the soil can become totally saturated during periods of heavy and prolonged rainfall. With high levels of natural runoff, often concentrated into a single channel, flooding associated with high volume stream flows (with the ability to transport large quantities of sediment) can be expected to occur on a periodic basis within many parts of the Asia Pacific region. It is worth remembering that the floodplains of the Indo-Gangetic river systems were developed by inundation from forest-covered mountains long before watershed damage by man had become a significant factor.

SARM in geologically recent hill and mountain landscapes has to recognise that various natural denudation processes are at work even in areas where there has been no human disturbance. Such processes have to be considered as natural hazards, and therefore fixed design constraints when seeking to develop improved land use management practices. As one recent UNESCO publication states "It is often conveniently forgotten that floods are a natural hazard in areas with heavy rainfall".

There is a distinction, although they overlap, between unsuitable land use and inappropriate land management practices. Unsuitable land use is the use of land for purposes for which it is bio-physically unsuited on a sustainable basis. An example would be the growing of annual crops on steep hillsides with shallow soils.

Inappropriate land management practices refer to the use of land in ways which could be sustainable if properly managed, but where the necessary practices are not adopted. For instance the failure to adopt soil conservation measures on sloping land or to replenish soil nutrients removed in harvested products. It can also refer to land use which is ecologically sustainable when the intensity of use is low due to an abundance of land but becomes inappropriate when land scarcity leads to higher intensity of use. Examples are shifting cultivation and the grazing of semi-arid rangelands.

Various types of human activity can be identified as direct causes of land degradation. These can be considered under the following headings:

— *Poor agricultural activities*: defined as the improper management of cultivated arable land. It includes a wide variety of practices, such as insufficient or excessive use of fertilisers, shortening of the fallow period in shifting cultivation, use of poor quality irrigation water, absence or bad maintenance of erosion control measures, untimely or too frequent use of heavy machinery, improper crop rotations etc. This category would also include the extension of cultivation onto lands of lower potential and/or high natural hazards. Degradation types commonly linked to this causative factor are erosion (water or wind), compaction, loss of nutrients, salinisation, pollution (by pesticides, fertilisers).

— *Deforestation and removal of natural vegetation*: defined as the near complete removal of natural vegetation (usually primary or secondary forest) from large stretches of land, for example by converting forest into agricultural land, large scale commercial forestry, road construction, urban development, etc. Deforestation often leads to erosion and loss of nutrients.

— *Overexploitation of vegetation for domestic use*: contrary to "deforestation and removal of natural vegetation", this causative factor does not necessarily involve the (near) complete removal of the "natural" vegetation, but rather a degeneration of the remaining vegetation, thus offering insufficient protection against erosion. It includes activities such as excessive gathering of fuelwood, fodder, (local) timber, etc.

— *Overgrazing*: besides actual overgrazing of the vegetation by livestock, other phenomena of excessive livestock amounts can be considered under this heading, such as trampling. The effect of overgrazing usually is soil compaction and/or a decrease of plant cover, both of which may in turn give rise to water or wind erosion.

— *Overexploitation of surface and groundwater resources*: In areas of non-saline groundwater, the technology of tubewells has led to abstraction of water in excess of natural recharge by rainfall and river seepage and a progressive lowering of the water table. In some parts of the region the over-extraction of water (for irrigation, urban and industrial use) from rivers and other surface water sources has led to reduced downstream availability. Where water is returned after use it may have a higher salt content and/or be polluted from agro/industrial-chemicals and human wastes.

— *Industrial activities*: includes all human activities of a (bio)industrial nature: industries, power generation, infrastructure and urbanization, waste handling, traffic, etc. It is most often linked to pollution of different kinds (either point source or diffuse).

The nature, extent and risk of land degradation, and the potential sustainable yield of individual crop, tree and livestock enterprises, will ultimately be determined by the prevailing biophysical conditions within a specific area. Decisions as to what their landholdings are actually used for, and the management practices to be followed, will be influenced primarily by the socio-economic circumstances in which individual rural households operate. While current land use enterprises and management practices may accelerate land degradation, technical remedies will only succeed if they can function within, and address, local socio-economic constraints.

In the past too much emphasis has been given to assessing what is happening rather than why it is happening. Priority has wrongly been given to tackling the visual symptoms of land degradation (e.g. soil erosion control, gully plugging, reforestation etc), whereas the first step should have been to analyse why undesirable land uses and poor management practices were being followed. Attention should be directed to identifying the ultimate cause, which in the case of accelerated (as opposed to natural) erosion, more often than not, will have a socio-economic origin.

Failure to consider the socio-economic dimension may result in the underlying causes of land degradation being overlooked and much time, effort and money spent in dealing with the symptoms of a problem rather than with the problem itself. SARM therefore requires that the issue be looked at, not just from a biophysical perspective, but in terms of the economic, social and political environment of those directly affected.

Downstream Sediments

Downstream sedimentation problems are attributed to erosion in the fields of small-scale farmers who grow annual crops within the upper reaches of a watershed. This is often backed up with experimental data from small research-managed runoff plots showing high rates of soil loss under so-called farmers' traditional practices. The true source of eroded material may lie elsewhere. What is needed is for more extensive field assessments to determine what the reality is on the ground.

For instance one of the highest annual rates of soil loss from one of the Fiji plots was only 2.6 tonnes per hectare, whereas an earlier study had suggested a rate of 60-66 tonnes/ha/yr. In Western Samoa infiltration rates were found to be high, often over 90% even on 25o slopes with over 4,000mm annual rainfall. The trial plot under the farmers' traditional practice produced virtually no soil loss or runoff. Such results beg the question as to the source of the observed downstream sedimentation problems in the country. If it is not from the fields of the small-scale farmers then it has to becoming from elsewhere, e.g. roadside cuttings, stream bank erosion or forestland.

Qualitative techniques exist for the visual field assessment of erosion and these should be used to determine whether erosion is taking place, and if so is it associated with:

— areas used for annual crop production (maize, upland rice, cassava, yams, taro, vegetables etc);

— areas used for perennial crop production (rubber, oil palm, coconuts, banana, coffee, sugarcane, fruit tree orchards etc);

— pastures and natural grazing areas;

— fallow areas;

— forest areas (plantations, woodlots, natural woodlands and logged over forests);

— settlement/factory areas;

— mining and quarrying areas;

— roads, access tracks and farm footpaths;

— unconsolidated and active land slide areas; and/or

— stream bank and river bed erosion.

In many cases the worst downstream sedimentation problems may be attributable to one or more of the following:

— Poor logging practices in conjunction with badly constructed forest extraction roads;

— Poorly built and aligned rural roads and farm access tracks;

— Uncontrolled expansion of settlements and industrial areas;

— Land clearing and development activities associated with the establishment of commercial mono-cropping (e.g. ginger in Fiji, cassava in Thailand personal observation) and orchards and woodlots on hillslopes (e.g. in Fujian Province PRC China personal observation). Thus the regions small-scale hillside farmers, who typically get the blame, may not be the real culprits.

Reversible Land Degradation

The effects of water and wind erosion are largely irreversible. Plant nutrients and soil organic matter may be replaced, but given the slow rate of natural soil formation, replacing the actual loss of soil material would require taking the soil out of use for many thousands of years, an impractical course of action. In other cases, land degradation is reversible: soils with reduced organic matter can be restored by adding plant residues, and degraded pastures may recover under improved range management. Salinised soils can be restored to productive use, although at a high cost, through salinity control and reclamation projects.

Land reclamation frequently requires inputs which are costly or labour-demanding or both. The reclamation projects in salinised and waterlogged irrigated areas demonstrate this fact clearly. In other cases, the land can only be restored by taking it out of productive use for some years, as in reclamation forestry. The cost of reclamation, or restoration to productive use, of degraded soils is invariably more than the cost of preventing degradation before it occurs.

9

Planning and Managing of Soil and Water Resources

In many developing countries, expanding populations exert increasing pressure on land and water largely in efforts to enhance food production, and with little regard to long term impacts. In the wealthier countries, human populations expanding in drylands because of favourable climates, but irrespective of water scarcity. In all cases, however, the wise use and management of soil and water resources are fundamental to achieving sustainable economic development.

Freshwater scarcity is a global problem that is particularly acute in drylands, and that threatens our ability to achieve food security, alleviate poverty, and improve human health. The occurrence of droughts compounds water scarcity and often leads to excessive use of natural resources. Extended droughts results in famine in the poorest of countries. In contrast, many drylands can experience periodic excessive rainfall that can cause flash floods and consequently, loss of life and property. With few exceptions, water scarcity, and the extremes of droughts and floods impact rich and poor countries alike.

Land scarcity compounds the problems of water scarcity, making people more vulnerable to the extremes of droughts and floods and leading to widespread exploitation of natural resources. Accelerated soil erosion can accompany intensive human development and natural resource exploitation, whether it is urbanisation or grazing and cropping on hillslopes. Surface and gully erosion exacerbate problems of low productivity, further diminishing soil resources which threatens the future productivity of the land. Water quality can become impaired, which when coupled with high sediment levels, constrains our ability to develop sustainable water resource management. Clearly, actions must be taken on many fronts to develop sustainable solutions and improved management of land and water in dryland environments.

Land use and water are inextricably linked together, but are not often managed in concert with one another. Watershed management offers the framework for achieving this integrated management approach to increase or sustain food and natural resource production while protecting the soil and water resources upon which this productive capacity depends. Implicit in this approach is the recognition that land use in uplands affects the flow and quality of water reaching downstream areas; in contrast, water resource development (e.g., reservoirs, canals, and others) can affect the type and intensity of land use throughout a watershed.

Transforming this recognition into effective solutions, however, is hampered by inadequate policies and an absence of institutions and organisational arrangements that are needed to achieve the integration and intersectoral cooperation upon which implementation depends.

Watershed Management

Nowhere is there a greater need for comprehensive watershed management than in drylands. Understanding and coping with the linkages between land use and water is critical. Water should be viewed as the most valuable product of the land and the one resource upon which all other production depends. Whether the setting is the southwestern United States, with its expanding urban population, or an agronomically based country in sub-Saharan Africa, the specific objectives of management can vary, but the overall goals encompass the production of goods and services with the underlying dependence on developing and sustaining water resources.

Hydrologic Characteristics

Drylands are characterised not only by low annual precipitation, but also by erratic precipitation that is subject to large temporal and spatial variability. While precipitation in many semiarid lands averages between 200-500 mm/yr, which is suitable for some crops, the extreme fluctuations in precipitation from year to year make such averages of little use.

In mountainous areas, orographic precipitation can result in high rainfall in upper elevations but with much lower precipitation in the valley bottoms. For example, annual rainfall in the Rif mountains of northern Morocco can reach 1,000 mm in the highest elevations, but annual rainfall is less than 250 mm a few kilometres away on the leeward side of the mountains. As a result, water yield enhancement strategies usually focus on higher elevation watersheds.

Precipitation alone does not define the climatic character of drylands. Aridity describes the normal conditions of drylands that are the result of factors that create low levels of moisture availability, which includes not only low annual precipitation amounts, but also high evaporation demands. Ratios of evapotranspiration to precipitation are generally large in drylands, exceeding 95% in some areas. As a result, soil moisture is normally low during much of the year and is usually the controlling factor affecting vegetative growth. The exception is in oases or other riparian areas, such as in the southwestern United States, where evapotranspiration can exceed precipitation amounts because of locally available soil moisture.

Whereas aridity is the normal condition of low available moisture of drylands, droughts result from periods of below normal precipitation that persist. Heathcote defines a drought as " . . . an unexpected shortage of available moisture sufficient to cause severe hardship human resource use . . . ," and points out that a drought can occur in any climate. Coping with droughts is particularly challenging in drylands, where water is normally of short supply.

As a result of low and erratic precipitation in drylands, ephemeral or intermittent streams are the norm rather than perennial streams. It is not uncommon for perennial streams to originate in higher elevation zones, or from springs or seeps at the toe of slopes in mountainous areas, and then become intermittent downstream from the source area. Transmission losses of water (i.e., water that infiltrates into the channel bottom) can however, result in a mounding of shallow groundwater below and near stream channels and in adjacent floodplains.

Dense vegetative communities in these riparian zones, often dominated by phreatophyte vegetation (water loving plants, often deep rooted trees), are indicators of such conditions. Saltcedar (Tamarisk spp.), Mesquite

(Prosopis spp.), and cottonwood (Populus spp.) are examples of phreatophytes which have been shown to transpire 1,000 mm to more than 2,000 mm of water per year from shallow water tables in the southwestern United States. During periods of prolonged rain, or snowmelt runoff from high elevation zones, the water table can rise to the surface channel, only to retreat below stream channel bottoms during dry periods as a result of excessive evapotranspiration by phreatophytes. As a result, perennial streams in downstream valleys are rare as rainfall or snowmelt cannot be relied upon to sustain streamflow throughout the year.

Floods are not uncommon in drylands and can occur during the rare periods when there is prolonged and excessive rainfall or snowmelt runoff, or more frequently, from convective rainfall that brings localised but intense rainfall. Flash floods from high intensity rainfall are highly variable across the landscape and are common in the southwestern United States. It is not uncommon for dry stream channels to become torrents within hours of convective storms that may have occurred several kilometres upstream. Furthermore, one stream channel can be flooding while another channel a few kilometres away can be completely dry. Such floods are the result of surface runoff or overland flow that occurs from watersheds with sparse vegetative cover and shallow, poorly developed soils with low infiltration capacities.

Much, if not all, of the stormflow from flash floods that results from runoff into dry channels is lost to transmission losses and bank storage. As a result, the volume of streamflow often diminishes rapidly as the stormflow event moves farther downstream. Some of the transmission losses may recharge groundwater aquifers, but much of the bank storage and water that recharges shallow groundwater may be eventually lost back to the atmosphere via transpiration by phreatophytes.

Sediment Delivery

Soil erosion reduces the productivity of upland watersheds and the resultant sediment that reaches stream channels can adversely impact reservoirs, water conveyance systems, and water quality in downstream riparian corridors. Soil loss by the actions of water and wind is generally higher in drylands such as the southwestern United States than in the more humid east. Surface soil erosion accounts for much of the soil loss in the southwestern United States, but gully erosion is important in heavily degraded areas.

Intensive grazing and cultivation of marginal farmland occurs in many drylands of the world and have caused excessive soil loss from surface and gully erosion. Gully erosion is an indicator of advanced watershed degradation and can reduce the production potential of the land dramatically. When accelerated erosion occurs immediately upstream of a reservoir, the resulting sediment delivered to a reservoir can impact the function and life of the reservoir. The following two examples are follows.

The small country of Lesotho in southern Africa has historically been intensively grazed which has caused accelerated levels of soil erosion. Surface and gully erosion rates in a 4.9 km^2 watershed in Lesotho were documented with aerial photographs and ground measurements. Gully eroded areas increased by 260,000 m^2 between 1951 and 1980. Over that period gully erosion caused (1) a roughly 5% reduction in productive land area of the watershed, and (2) significant contributions of sediment to a downstream reservoir. The sediment delivered to a downstream reservoir was 300,000 t and 80,000 t, respectively from gully and surface soil erosion. The trapped sediment within the reservoir had a corresponding volume of about 267,000 m^3.

Mean annual precipitation averages less than 250 mm/yr in Morocco, but ranges from less than 50 mm/yr in the far south and southeast to more than 1,000 mm/yr in the highest elevations of the Rif, and the Middle and High Altas mountains. Nearly 90% of the country receives less than 250 mm/yr. Because of this uneven distribution of precipitation, reservoirs have been constructed near these high elevation zones to provide irrigation water for the drier areas. Existing and planned reservoir projects, however, are threatened by excessive soil erosion in many parts of Morocco.

In addition to the impacts on reservoirs, excessive soil erosion in many parts of Morocco is causing the productivity of soils to decline. Annual loss of soil nutrients in agricultural areas located northwest of the Rif mountains is equivalent to $US20/ha/yr. This represents about 90% to 180% of what local farmers invest yearly in fertilisers.

In the Lesotho case, once gully erosion begins, and if land use and structural changes are not implemented early on, land degradation proceeds rapidly with the potential for serious economic impact both on-site and downstream. The implications for watershed management would suggest that erosion from overgrazing led to both a loss of land production potential and shortened the economic life of the reservoir.

Costly rehabilitation efforts involving both structural and vegetative measures would be needed to reverse these processes. From a watershed management perspective, preventing such losses in the first place through improved grazing practices can be (and should be) considered as an economic benefit, and thus a basis for justifying grazing management programmes.

In the Moroccan example, land use and rainfall interactions dictate soil erosion and sedimentation rates. The highest rates of sedimentation correspond to the watershed with the most intense land use (and least forest cover), but with the lowest rainfall, averaging 350 mm/yr. Herein lies a dilemma for watershed management. With such low rainfall and severely overgrazed watersheds, opportunities to reduce erosion (estimated to be >250 t/ha/yr) and sedimentation through re-vegetation are limited. Costly structures would be required to accompany any vegetation programme which takes a long time to materialise; furthermore, the ability to reduce grazing pressures during the rehabilitation period dramatically constrains rehabilitation in such areas where the rural poor are directly dependent on their livestock.

As a result, reservoirs fill with sediment at an alarming rate. Where rainfall is greater, the potential is also greater for restoring the watershed. A watershed restoration programme that controlled grazing and increased forest and fruit tree production on the Loukos watershed was shown to be economically justified with an internal rate of return of 15.9%. There is another aspect to sediment delivery at reservoir sites that needs to be addressed when considering watershed restoration measures. Sediment that is entrapped in streamflow is the product of upland erosion processes, streambank erosion, and channel scour.

Streams are formed over geologic time in response to dynamic and complex fluvial processes and hydraulic factors. When watershed conditions change sufficiently to alter either the streamflow response or the amount and type of sediment that enters the channel, the stream will adjust and can become unstable, in some instances accelerating sediment delivery downstream. Similarly, changes in the riparian plant community that alters streambank stability and that affect the velocity of flow through floodplains during flood events, all affect downstream flows and sediment regimes. Therefore, floods and sediment are linked together and are further linked to the condition of upland watersheds and their riparian plant communities. Flash floods are natural, episodic events that transport large quantities of sediment and debris. The extent to which land use on upland watersheds affects the magnitude and frequency of flash flooding, and the sediment and debris load that these floods carry, is not well understood in most drylands.

Challenges of Ecosystem Management

Pressures on natural resources are intense in the drylands of the world. As populations of people and livestock increase, their demands for food, forage and other natural resources increase, thereby applying stress to drylands. Responding to such stresses and managing these ecosystems is constrained by inherent characteristics of sparse vegetation, shallow soils, low and undependable amounts of precipitation, and generally low productivity. In the developing world, overuse of drylands is expressed through overgrazing, depletion of fuelwood, and cultivation practices that are not suited for the climate or terrain.

Even in the developed world, considerable stress is applied to the land through overgrazing, urban expansion, excessive use of off-road vehicles, and so forth. In all drylands receiving intensive use, some common issues and problems arise, which need to be addressed through management. Foremost, drylands have commonly been viewed in the past as wastelands, not worthy of economic concern or political attention. Global concerns about desertification over the past few decades have focused more attention to dryland issues and the need for land use reform.

Desertification is often used to describe areas that have become desertlike in their appearance as a result of human induced degradation. To some extent, emerging programmes to combat desertification have helped generate the political and economic support that is needed to reverse land degradation. However, there are certain inherent characteristics of drylands that place limits on the potential for agricultural, natural resource and urban development. There are also conditions that make watersheds vulnerable to degradation, and that constrain our ability to restore or rehabilitate the land.

Water scarcity is a reality that must be dealt with before any type of development can be sustained. Opportunities for enhancing water supplies are discussed later, but there are other factors that need to be addressed in the planning and management of land and water resources so that the productivity of the land is sustained, and the soils and vegetation communities of dryland watersheds are not degraded. Any type of development of drylands should recognise that vegetative communities and soils are sensitive to intensive use and once degraded, many decades are needed to restore the production and hydrologic function of these ecosystems.

Low and erratic precipitation prevents the rapid reestablishment of vegetation, leaving a degraded landscape exposed to water and wind erosion for long periods of time. Although aridity typifies the general condition of drylands, it should also be noted that most dryland regions have "islands of abundant moisture" that can be found in higher elevation mountains (as in the southwestern United States), in oases, and in other riparian areas. Because of the presence of water, high productivity, and ecological values, riparian areas receive an inordinate concentration of human and animal use, many times leading to overuse and degradation. As a result, riparian areas have become a focal point of natural resource managers and developers in drylands.

Because of the favourable climate, many drylands are targeted for urban and agricultural development, all of which hinges upon the ability to develop and enhance water supplies. Many large cities and retirement communities have emerged in the driest areas of the southwestern United States. In contrast, large-scale irrigation projects likewise have emerged in the drylands of China, India and Pakistan. The sustainability of such efforts is becoming problematic as populations grow, demands for water increase, and conflicts over water use arise. In the past, engineering technologies have been relied upon that capture water in water abundant areas (or during periods of high precipitation) and then store and transfer water to areas (or during time periods) where or when it is needed.

Water Supplies of Drylands

An array of technologies exists to increase water supplies in an area, including cloud seeding, towing polar icebergs, desalinisation of sea water, groundwater development, reservoir storage, evaporation suppression and vegetation management. New technologies are emerging but most are costly and sometimes have unwanted environmental impacts. A full discussion of all these technologies is beyond the scope of this paper. What is stressed here is the need for comprehensive to develop effective and efficient water resource development projects within the framework of sound watershed management. There is no simple solution. The management of uplands and riparian areas must be coordinated among stakeholders in a way that makes good economic and ecological sense. The following methods can be considered as complementary to the more the more common and larger scale reservoir storage and transfer technologies

Vegetation Management

Vegetation can be managed to either conserve existing water supplies or to increase the yield of water from a watershed for a given precipitation regime. Conservation of existing water supplies can be achieved through more efficient irrigation methods and by favouring crops with inherently low consumptive use. The selection of crops that have low water requirements will depend on the region and species that are adapted to the local conditions. Of course, any such crops must be acceptable to local people and must be economically viable. Therefore, a detailed discussion of specific species is not possible. In most areas, we do know that certain types of vegetation use water extravagantly and should be avoided. Eucalyptus trees are a good example of rapidly growing trees that have been planted in many drylands for purposes of increasing fuelwood supplies.

The cost of growing such trees needs to be considered in terms of water consumption, that is, water that is lost that can otherwise be used for other purposes. It is not always clear whether any water savings from altering species will actually be realised. Genetic engineering may eventually lead to the development of crops with low consumptive use, although at this time the authors are unaware of any such research that has produced such plants. Manipulating vegetation on watersheds to increase water yields has been given considerable attention for several decades.

Studies indicate that water yield can be increased when: (1) forest cover is reduced, (2) deep-rooted species are replaced with shallow rooted species, and (3) species with high interception and transpiration (ET) losses are replaced with species that have lower ET losses. Hibbert considered opportunities for increasing water yield from drylands in the western United States and concluded that the above manipulations of vegetation increase water yield only when annual precipitation exceeds 400 mm/yr, but increases are marginal unless annual precipitation exceeds 500 mm/yr.

Before vegetation manipulation for water yield enhancement is implemented, several factors must be taken into consideration. First, there must be some means of capturing the water yield increase; reservoir storage should be both adequate to accommodate any increases in water yields and close to the watershed outlet. If water yield is increased during periods when reservoirs are full, there is little value in carrying out such management.

Changing vegetative cover on a watershed can affect wildlife habitat as well as other uses of the land, such as outdoor recreation. Changes in erosion, sediment transport, and water quality must all be understood prior to implementation. Vegetation manipulations for purposes of conserving groundwater were the focus of early work that considered the removal of phreatophytes in floodplains, or otherwise reducing their transpiration. Because of the high transpiration rates of phreatophytes is considerable potential to salvage groundwater in areas where groundwater supplies are critical.

Such possible benefits must be weighed against the value of such vegetation in terms of riparian values for flood control, channel stability, wildlife habitat, and so forth. Clearly there are limitations in many dryland areas to enhance water yield through vegetation manipulations. Other technologies can be used to increase water yield or to complement other water conservation measures. Some are tailored more for local household or farm use, while others can be components of more comprehensive water management systems.

Water Harvesting

For centuries people have captured rainfall from rooftops that is stored in barrels for later household use. Likewise, but on a larger scale, small impervious catchments have been designed to efficiently produce runoff from rainfall as a means of increasing water supplies in drylands. Storage of the harvested rainfall is required, and such systems are normally used to augment existing water supplies or as an alternative to more costly methods of obtaining water. Water harvesting systems are small scale and best suited for watering livestock, small-scale subsistence farming, and/or local domestic use. Some water harvesting systems have been used for centuries and still have application in certain situations.

Several different designs are possible, but generally at least 80-100 mm of annual rainfall is required. Apron type water harvesting systems have an impervious catchment area, a storage facility and a water distribution system and have been widely used for providing water for livestock. Catchment areas can be a few square meters up to 1,000 m^2, and consist of natural rock outcrops, paved roads, or soils that have been treated to make them impervious. For livestock watering, storage tanks should be covered or situated where evaporation losses are minimal.

Water harvesting has been widely practiced in sub-Saharan Africa. Such systems have been effective in augmenting rain-fed agriculture in areas with unreliable and erratic rainfall that normally occurs over short periods of 70-120 days per year. For example, in Burkina Faso and Kenya, catchments of 1-1.2 ha have been used to harvest rainfall to irrigate crops where annual rainfall averages 600-700 mm/yr, but potential evapotranspiration exceeds 1,500 mm/yr. Such applications can be crucial to subsistence farmers and pastoralists in many parts of the world. Economics, social factors, and careful consideration of alternatives should be taken into account before investing in water harvesting systems.

Water Spreading

Ephemeral flows of streams in drylands can sometimes be diverted onto gently sloping areas to enhance forage production or to irrigate crops in drylands. Water spreading systems are best suited where upstream watersheds produce at least three or four runoff events immediately before or during the growing season. Stormflow is diverted from a channel and then with the use of a system of long, low earthen dams or dikes that are parallel to the contours of the land, water is spread onto a gently sloping land surface. The system is designed so that a series of cascading dikes splits the water received from upslope, so that water is spread over an increasing area in the down slope direction. The diverter can conduct all or just a portion of streamflow directly into the spreading system. Even more so than with water harvesting, water spreading systems have limited applications.

Although water scarcity is pervasive in drylands of the world, and has been the focus of the preceding discussion, the role of watershed management in coping with the extremes of droughts and floods in drylands also deserves special attention. Droughts and floods are often referred to as unexpected and rare, and crisis management seems to be the term that best describes how many parts of the world cope with such occurrences.

In planning for land and water management and human development in general, we should, however, recognise that such events do occur, and in fact they are not necessarily rare. What may be rare and unexpected is their magnitude and duration, not to mention the rarity of preparedness on the part of people affected by their occurrence. The weak links in actually coping with droughts and floods, however, are usually related to poor planning, the absence of a comprehensive, integrated approach to land and water management, and the means to coordinate and implement comprehensive programmes.

To begin with, planners and managers must recognise that droughts and floods will occur, and that people can do little to affect their occurrence. However, we can affect their impact on humans. Land and water scarcity affects human behaviour and to some extent constrains our ability to cope with extreme hydrologic events. In most drylands, people become concentrated near stream channels, and in the adjacent riparian or floodplain areas. Such areas are, of course, the most prone to flooding.

Even though floods may not occur every year in ephemeral streams, we know that they will eventually occur and the nature of flash floods gives little time for humans to escape their destructive power. In mountainous terrain, steep stream channels can become charged with sediment over time, which can become debris torrents when high intensity rains occur. Such debris torrents are extremely destructive to the downstream areas that lie in their path. One or more of the following approaches can be taken to cope with floods and debris torrents:

1) flood and debris torrent control with the use of engineering structures such as dams, levees and diversions,
2) flood warning programmes that give people advanced warning of an impending flood so that they can leave the flood prone areas, and
3) zoning of hazardous areas that restricts human activity in areas susceptible to floods and debris torrents.

Given that the first option is costly and can provide only limited protection, and flood warning systems in areas subjected to flash flooding is unrealistic, zoning of hazardous areas (e.g., flood plain management) is considered to be the most viable option in most cases. This approach recognises that where people and their dwellings become concentrated in areas that are prone to hazards of flooding or debris flows, disasters will occur.

Policies and institutions must be in place that provides incentives for people to avoid hazardous areas. Delineating flood-prone areas can be facilitated with GIS-based terrain analysis. Similarly, methods of delineating flood plains and zoning are well known. However, incentives are needed to change people's behaviour in such hazardous areas; an example is the Federal Flood Insurance programme (USA) which links insurance rates to the degree of hazard.

Coping with droughts is more problematic and requires that many contingencies be considered. Two perspectives are needed, one to prepare for droughts and the other to deal with droughts after they occur. Efforts to increase water supplies in drylands and that can carry over supplies for several years have been the goal of many water resource projects. Reservoirs, water transfer systems, etc. all have some merit in developing dependable water supplies for periods of water shortages.

Once droughts occur, water conservation programmes can become accelerated by means of reducing consumption. However, communities that already have strict water conservation programmes in place may find that once the drought occurs, there may be no buffer in the system; that is, there are no additional water conservation actions that can be implemented without incurring severe hardship. In many instances, programmes can be established that incrementally reduces water use on an established priority basis at the outset of droughts.

Restrictions on water use can become prioritised based on established criteria for a region. Those uses, which have the least adverse impact on people, become restricted early on and progress into uses that have greater impacts. In any event, coping with droughts requires the necessary policies and institutions that can implement a combination of water conservation measures and drought contingencies. Options, unfortunately, seem to be more restricted in the poorer countries, where the effects of droughts are more dramatic.

A watershed management approach can play an effective role in coping with the types of hydrometeorological extremes discussed above. In the process of planning and implementing agricultural, water resource, and other natural resource projects, a watershed approach brings into the process the explicit recognition of the linkages that exist between land use and water and between upstream and downstream areas. Land use that leads to loss of vegetative cover, soil disturbance and overall watershed degradation can exacerbate flooding and debris flows and can constrain the ability of local inhabitants to cope with droughts. The benefits of soil and water conservation through watershed management are expanded upon below.

Issues and Challenges in Watershed Management

Dryland watersheds present unique problems and challenges to planners and mangers of land and water resources. They are more fragile than watersheds with more abundant rainfall and dense vegetation. They have a limited potential for production that is largely due to inadequate precipitation and poor soil conditions. Once disturbed, these watersheds take longer to recover. For the most part, water scarcity dictates the types and extent of land use that takes place, yet we see land development occurring in some drylands as if water was abundant.

In moving toward sustainable development and use of resources on these lands, the biophysical realities of these systems must be recognised, but equally important, the necessary policies and institutional arrangements must be in place to mobilise human activities on the land in ways that are compatible with these biophysical realities.

Freshwater benefits to downstream areas naturally accompany sound management of upland and riparian areas. Preventing watershed degradation should, therefore, be realised as beneficial; avoiding losses that would occur from exploitation of resources on the land is equally beneficial as increasing the productivity of an already degraded land. In contrast, management can be targeted to specific watershed objectives, such as enhancing the quantity and quality of water yield above a reservoir. In either case, benefits can be masked by: (1) the location and diffuse nature of land use practices and their effects, (2) the scale of activities relative to overall watershed size, and (3) the length of time needed for benefits to become realised.

Changes on the land can have incremental effects, which individually may not be apparent, but when viewed over the watershed and over time, can have more serious cumulative effects. This complexity has blurred the vision of decision-makers in many parts of the world, has constrained economic assessments, and overall weakened commitments to watershed management. To overcome this problem, the economic costs and benefits of comprehensive watershed management must become better known.

Watershed improvements for land and water management purposes require economic justification. A watershed perspective provides clarity in determining the economic importance of water-related ecosystem services (e.g., increasing water yield, improving water quality, reducing sediment delivery to a reservoir) attributed to changes in land use. To date, comprehensive economic analyses that consider the full range of these services or benefits from watershed improvements are lacking.

Policy Considerations

More effective and efficient management of land and water resources to improve human welfare requires more than just technical knowledge. Technical information provides a foundation from which upstream-downstream linkages can be assessed and economic analyses performed. Transforming such information into management requires effective participation by stakeholders to develop a consensus and provide incentives for implementation.

A policy environment must be created that supports rather than hinders the integration of land and water management. Watershed and political boundaries rarely coincide; as a result, the necessary coordination of land and water management depends upon functional organisations that can resolve transboundary issues and water use disputes. The lack of effective watershed-level organisations led to the formation of more than 1,000 watershed districts in the United States during the 1990s to deal with upstream-downstream issues.

Internationally, the Nile Basin Initiative established a basin-wide partnership of nine riparian countries to help resolve transboundary issues and to deal with inequities of water distribution in the basin. More than 80% of the flow in the Nile River originates in mountainous Ethiopia, a river upon which downstream Sudan and Egypt are heavily dependent. Without cooperation and coordination of both land use on watersheds and water development projects within the basin, disputes and conflicts could erupt. In all situations, a strategy is needed that promotes coordination and cooperation among the stakeholders in a watershed.

10

Soil and Water Resource Management

Agricultural productivity requires the economically appropriate use of germplasm, crop protection techniques, and also the careful management of soil, water and nutrient inputs. For best results, this necessarily includes protection of the on-site natural resources, so productivity and on-site short-term sustainability have always run in parallel where a farmer has a long-term interest in his land. However, the concept of sustainability has now expanded in terms of both space and time.

All consequential environmental damage, including off-site or even global effects, is to be controlled, and the maintenance of the resources must be considered over long timespans, certainly into the next generation. There is a recognition that all these conditions must be in place to allow truly productive and sustainable agriculture to be carried out, whilst bearing in mind that the underlying purpose is to improve food security for the poor, at a time when the population of the less-developed countries is increasing at the rate of about 1 billion per decade.

Degradation of natural resources, in particular of the soil, is proceeding at an increasing rate. There are several causes leading to different types of degradation, but a particularly pervasive underlying cause is poverty. The poor are to be found in many parts of the world but, overwhelmingly, they are still located in rural areas of the less-developed world. These people are directly dependent on rural natural resources, of which they themselves are the key managers-users, custodians and exploiters. Their poverty, aided and abetted by their insecure tenure, frequently obliges them to take an extremely short-run view of their custodial responsibilities, and their decisions individually and collectively often result in their resources becoming degraded, sometimes in dramatic and socially unacceptable ways.

Land and water (as well as forest and fishery) degradation is thus importantly driven by rural poverty, in turn determined by many factors, ranging from population growth and rural-urban migration, to rural-industry diversification, infrastructural development and employment, as well as resource degradation itself. Interventions, including research-based technology and policy improvements, that assist in alleviating poverty or increasing sustainability will thus have strongly synergetic effects and are best seen as necessarily complementary approaches.

For environmentally sustainable agriculture, the whole resource base has to be protected and enhanced. Within this, the CGIAR paper "A Research Agenda for the Future" of 6/10/94 noted the increasing importance of efficient water use, and stated five major research thrusts for the CGIAR System: water and irrigation management; "ecosystem (including watershed and river basins) management; ecological foundations of sustainable production (including soil/water/nutrient/plant/animal relations); ecosystem conservation and restoration; and common property resources." More recently, the February 1995 Lucerne Action Plan requests CGIAR to "Address more forcefully the international issues of water scarcity, soil and nutrient management, and aquatic resources."

Various recent initiatives are in progress or in planning within the CGIAR System that contain an emphasis upon soil and water. This includes work under most of the Ecoregional and Systemwide programmes which have started in the last few years. The Soil-Water-Nutrient (SWNM) Programme is now being planned by a group from CGIAR and other institutions, and was printed as the - "Zschortau Plan".

The sustainability of natural resources includes many environmental issues, in which CGIAR will have an interest. However, it is essential that this review should maintain a clear focus, and TAC has specified that this study should concentrate most on soil and water as explained in the main report; to quote the Framework paper TAC which was prepared to guide the study: "TAC decided to limit the present study to terrestrial ecosystems, and more specifically focus on the use and conservation of soil and water resources for the sustainable benefit of humans, particularly through the contribution of these resources to sustainable agriculture and livestock production." In practice, this requires the inclusion of plant nutrients, because of the role of the soil in supplying them.

This requires consideration both of the practical use of these resources in the production of food and fibre, and the consequences of that use for the quality of the resources. Any damage to the resources then impacts upon both the further production of food, and on other environmental and sustainability issues. Some of these are directly relevant to the CGIAR, such as water quantity and quality for water supply, fish farming, agricultural biodiversity, forestry and marine impacts, though they are not main subjects of the study.

A number of other issues may also be touched upon, which are not close to the interests of the CGIAR, but which are regarded as important in that they were identified in Agenda 21 as related to natural resources management (NRM). This includes general biodiversity, biogeochemical cycling processes, landscape value, leisure uses and processes that are important for global change. As noted in the main report, NRM with a strong emphasis on sustainability can appropriately be defined as Integrated Natural Resource Management (INRM).

Sustainability and Productivity

In "A Research Agenda for the Future" the CGIAR stated that agricultural outputs would have to be doubled in the next 30 years, and that this has to be done without damage to the environment. Following this, the CGIAR appointed a Task Force to study Sustainable Agriculture. There are many definitions of sustainability, but the following definition is used here, as it agrees well with that used by the Task Force. It is also close to that adopted by Smyth and Dumanski in considering the evaluation of sustainable land management. A sustainable agriculture must be:

(a) economically viable;

(b) must aim to protect its supporting natural resources;

(c) must be broadly acceptable to the population at large, including all off-site problems.

Economic viability is obviously essential, but it is difficult to generalise about because it is so open to change through market instability and government policy, sometimes on a very short-term basis. Soil salinity is generally accepted as being highly damaging, but it becomes much less so if it were intended to grow only a fully salt-tolerant crop variety. Acceptability to the population covers a great range of factors, from religious beliefs to health concerns, animal welfare and popular environmental issues, all of which are often country- or region-specific.

It is obvious from the above definition that all socioeconomic and biophysical factors must be of critical importance for Natural Resource Management (NRM) research. Interference with the world's biogeochemical cycles, such as net carbon flow to the atmosphere or nitrogen flow to the sea must be controlled. Impacts on biodiversity, measured on both regional and global scales of importance, must be minimised. Decisions about how strict these standards must be is very difficult to take, but progress is being made.

"Sustainable development" was made popular by the Brundtland Commission. It did not, however, define how much development is sustainable, or for how long. No one can foresee what will be considered essential after a long time span. Sustainability is not an absolute, but varies with time and place, and each case needs careful analysis. For example, improving agricultural productivity has a general value in preventing environmental damage, in that it lessens the pressure for the development of more fragile land and water resources. In this way intensification can actually improve over-all sustainability.

An appropriate precautionary approach in natural resource management is, where possible, to prevent degradation and pollution rather than to remedy it. It is unavoidable that damage will occur in some cases, and much attention must be given to the linked questions of reversibility and resilience. If any environmental damage that is caused can be reversed relatively easily by stopping or reversing the action causing it, then some degree of damage may be acceptable for a period. But if the damage is irreversible, such as the total extinction of a species, or massive erosion of a mountainside, then all efforts should be directed towards immediate prevention.

Reversibility will depend upon the type of action, its intensity and duration, and the resilience of the natural resource acted upon. Resilience implies the ability to return to its initial state after being stressed. The concept is more complex than appears at first, because resilience is not a single value. For example, the water quality of a river may show high resilience, because as soon as the pollutant source is removed, the river water returns to a pure state. However, the physicochemical resilience may not be matched by the biological resilience, because the river biota may have been destroyed during the pollutant incident. These various forms and degrees of complexity in the concepts of reversibility and resilience require a careful and critical approach, because many subjective judgements are involved.

There is a rather widespread belief that high productivity and sustainability are in some way incompatible. However, it is easy to find examples to show that this is not so: both low and high productivity systems can be both sustainable and non-sustainable. Thus over-grazing of hillsides is both unproductive and unsustainable; traditional long-fallow shifting cultivation is unproductive but sustainable; high intensity rice or wheat with careful agronomy is both productive and sustainable, whereas careless use of irrigated land is productive but unsustainable. Sustainability always demands care, and the higher the productivity needed, the greater the care demanded, and this is a problem that research has to solve.

Land Use

A holistic approach to natural resource use, such as is involved in a combined demand for productivity and sustainability, must involve the concept of land use and its allocation. The mosaic of topography, soil and water within a landscape, watershed or region can be used in many ways. These will have a scale of priorities in socioeconomic terms, by being in greater demand or producing a greater profit, but they also have to be tested against biophysical principles. The problems that can arise are best seen in the pressing demand for farm land in some less-developed countries, which leads to the use of almost any land, however unsuitable.

Some 250 million people in the tropics still depend on some form of shifting cultivation, which is gradually becoming less and less sustainable as the fallow periods decrease. The question of which form of land use will succeed this is of extreme importance. Unsuitable land uses will in the medium to long term prove to be unsustainable, and the sooner this can be determined, the less damage is likely to be done.

It is a relevant question how much work on land capability research that the CGIAR Centres should do. There is still a strong need for better biophysical research underpinning of the physical classification of land and water resources, especially exploiting the rapidly growing synthesis opportunities afforded by GIS methods and cheapened microcomputer databases. Demographic and socioeconomic information is also critical to such work, and the data and trends that drive them are rather local and national in scope and nature. Indeed, the responsibilities for such knowledge assembly and interpretation normally lie within national planning agencies, and execution within local and national governments.

Causes of Global Change

Most medium- and long-term planning of agricultural research in the past has tacitly assumed that the state of the atmosphere and the climate will remain constant, and most of our assumptions about agricultural sustainability depend upon this. This can no longer be taken for granted, because of the anthropogenically caused global change, the three drivers for which are land use change, atmospheric change and climatic change. The increase in concentration of carbon dioxide in the atmosphere is a proven fact, and it will continue to increase for many years, even on optimistic assumptions. The direct effects are likely to be benign, with increased growth rates likely in many plant species, but there are many uncertainties about the impacts.

The main dangers lie in climatic changes, in mean values or in variability. Such changes could well start to occur within the next 20 years, and even small changes could cause serious agricultural problems, at a time when the rate of population increase is maximal. No specific actions are possible at present, but precautionary planning should include these possibilities.

CGIAR would have a vital role in regard to climate change impacts in the tropics if and when it happens. Several mathematical models have already been constructed to predict the effects of climatic and atmospheric change. These have generally concluded that the less developed countries in the tropics would have the greatest problems in dealing with these changes.

Watershed Approach

Watershed approach is not relevant in all situations, and good research may be done without invoking it. Nevertheless, most land is part of an identifiable watershed and, depending on the size of the watershed, the concept can give a deeper and clearer perspective on sustainable agriculture, and soil, water and land use. The basic natural resources for agriculture are climate, geology, landform (topography) and the soils that develop from them, together with the biota that contributed to soils formation, and those (and their genetic resources) that are used for agricultural purposes. Soils and landform develop together, and in many parts of the tropics both are very old.

A watershed is a very appropriate way of looking at these resources within ecoregions, because it shows the spatial linkages very clearly. It is the logical unit for consideration of hydrological questions, including water demand and supply, irrigation, water quality and quantity, because the total size of the water resource initially available in the watershed can be precisely defined. A watershed will in effect be a cross-section or sample of the whole landscape. It will therefore include examples of all major landforms, apart from the overall constraint of the total altitudinal range within the watershed. For similar reasons, it will include examples of most major soil types found in that climate, subject to the distribution of rainfall across the watershed.

In terms of landform and soils, a watershed is therefore likely to be a very heterogeneous unit. Its value lies precisely in providing a sample of the total landscape, rather than a homogeneous block, when the interactions between various parts of that landscape are to be investigated. For similar reasons Greenland et al. classify the spatial dimension for cropping system research as the field plot, for farming system research as the field-village, but for NRM research as the watershed-region.

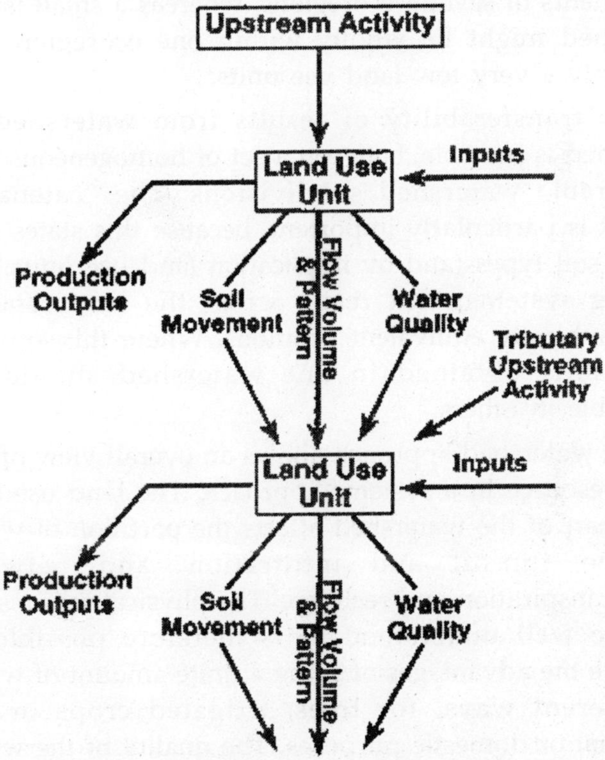

Figure 1. Land Use Units - Linkages in a Watershed

The further subdivision of a watershed is therefore a very important step in the planning of research, in which topography, climate, geology, soil type and farming systems will all be taken into account. It is therefore necessary to define the "land use unit": an area within which the climate, soil, farming systems, land cover and socioeconomic conditions are so homogeneous that any solution to a land use problem is probably applicable over the whole unit, and similarly defined units elsewhere (Figure 1). A typical very large watershed would include many different land use units, and indeed components of several ecoregions, whereas a small (sub)-watershed might lie wholly within one ecoregion and have only a very few land use units.

The transferability of results from watershed to watershed is possible, between a set of homogeneous and comparable watershed subdivisions. The "catenary" concept is particularly important, because this states that certain soil types (and by implication land use units and farming systems) will recur across the landscape in topographically equivalent positions. Where this applies, the results obtained in one watershed should be applicable in others.

The watershed approach allows an overall view of the water resource in a systems approach. The land use in a given part of the watershed affects the partition of water between run-off and infiltration, and between evapotranspiration and recharge. The physical reasons for this are well understood. It is therefore possible to compare the advantages of using a finite amount of water in different ways, for trees, irrigated crops or for industrial or domestic purposes. The quality of the water in different parts of the watershed can also be monitored and to some extent controlled. This is important for health reasons, e.g., it is believed that up to 3.5 million children die each year from water-borne diseases.

The focus on watersheds in relation to land use is by no means new, though few if any earlier projects have attempted the wide scope of the work discussed here.

Recent studies have included Lal and Russell, Pereira and Gregersen et al. The earlier East African work is particularly relevant. Pereira suggested that CGIAR should strongly emphasise this approach. The interest in water supply and use within the CGIAR System at that time ultimately led to the setting up of IIMI by a group of donors, but the focus on the watershed did not materialise. There appears to be only one significant watershed experiment in the CGIAR System at the moment, namely the relatively small SCOR project run by IIMI in Sri Lanka.

11

Managing Soil Erosion

Catchments in Soil Erosion

The catchments represent a resource management domain with biophysical and socioeconomic characteristics common to the marginal sloping uplands. The tropical parts of Asia suffer from serious soil erosion caused by intense rainfall that occurs within a few months of the year, and the topography of cultivated upper catchments (Figure 1). With nearly 60% of the total land surface covered by mountains and high plateaus, soil erosion remains a major problem to sustainable agricultural development in these marginal uplands. In addition to reducing agricultural productivity on-site, erosion causes negative impacts off-site.

The challenge is to recognise that land degradation, including soil erosion, is driven largely by socioeconomic factors. Research on soil erosion has largely ignored the end-users, or has been based on small plot experiments of no relevance to the watershed situation. A new paradigm for research is required which ensures that the whole range of stakeholders, from land users to policy makers is involved in the generation and promotion of improved land use practices.

Figure 1. Soil erosion in Asia is severe because of intense rainfall and steep topography of cultivated upper catchments

In Australia, Europe and North America, integrated catchment management has been adopted widely as the most effective way to involve stakeholders, and to utilise scientific knowledge in the management of land and water resources. However, there is still a strong need for capacity development of the NARES, especially in Asia, which will be involved in this kind of research. This has led to the establishment of the Management of Soil Erosion Consortium (MSEC), which is one of four consortia in the Soil, Water and Nutrient Management (SWNM) Systemwide initiative of the CGIAR.

In 1998, MSEC initiated a research project on soil erosion management in seven countries in Asia (Indonesia, Laos, Nepal, Sri Lanka, Philippines, Thailand and Vietnam) with support from the Asian Development Bank (ADB) under ADB-RETA 5803 "Catchment Approach to Managing Soil Erosion in Asia". The project in Sri Lanka was not continued because of implementation problems.

IBSRAM served as the Executing Agency until April 2001 when IWMI established the Southeast Asia Regional Office in Bangkok, Thailand, and took over supervision of the project. The project aimed to develop and promote sustainable and socially acceptable community-based land management options for sloping uplands through a participatory and interdisciplinary approach at a catchment level. The specific objectives are:

1 Develop sustainable and acceptable community-based land management options within a catchment framework;

2 Quantify and evaluate the biophysical, environmental and socio-economic on- and off-site impacts of soil erosion;

3. Generate reliable information and scientifically-based guidelines to improve catchment management policies;

4. Enhance the capacity of NARES on integrated catchment management and soil erosion control.

The objectives have been set to address the important issues as perceived by the major stakeholders. The design process was carried out with their active involvement.

An organisational model that engages scientists and research institutions to tackle a common goal was employed. The model allows the participation of those who can contribute, exploits synergies, and is mutually beneficial. Research planning was undertaken through consultation among concerned NARES, IARCs, NGOs and farmers. A facilitator, a steering committee and an annual assembly were essential to ensure the effective operation of the consortium. The NARES play the central role in the consortium, particularly in participatory research, but with a broad responsibility for underpinning applied and strategic research (Figures 2 and 3).

IWMI serves as the consortium facilitator, and together with other international centres and advanced research institutes provides the scientific inputs in the strategic and basic research. The whole idea of the programme is to take a bottom-up approach in research planning with iterative discussions between farmers, NARES, IARCs and ARIs in the definition and implementation of the research undertaking.

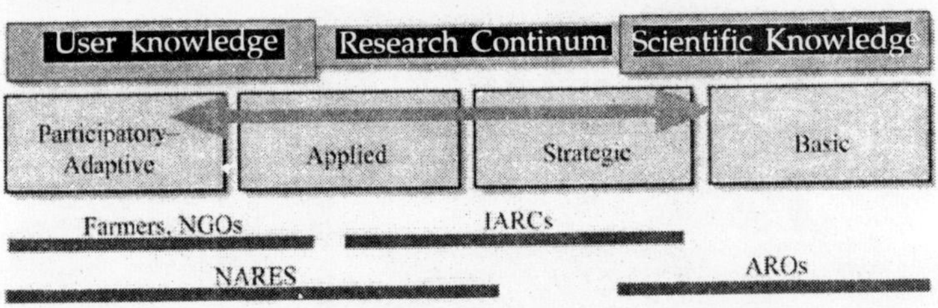

Figure 2. The research continuum showing the role of different groups in the implementation of MSEC research

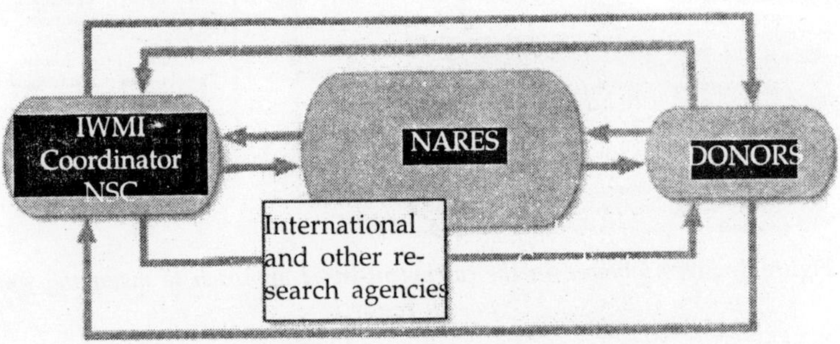

Figure 3. Schematic organisation of a network/consortium

The overall procedure followed by the consortium in programme planning and implementation is shown in Figure 4. Identification and selection of representative catchments in participating countries was conducted by an interdisciplinary team using carefully defined criteria and methodological guidelines, based on visits and dialogues with local institutions, scientists and farmers. These catchments were equipped with proper instrumentation for data collection and monitoring.

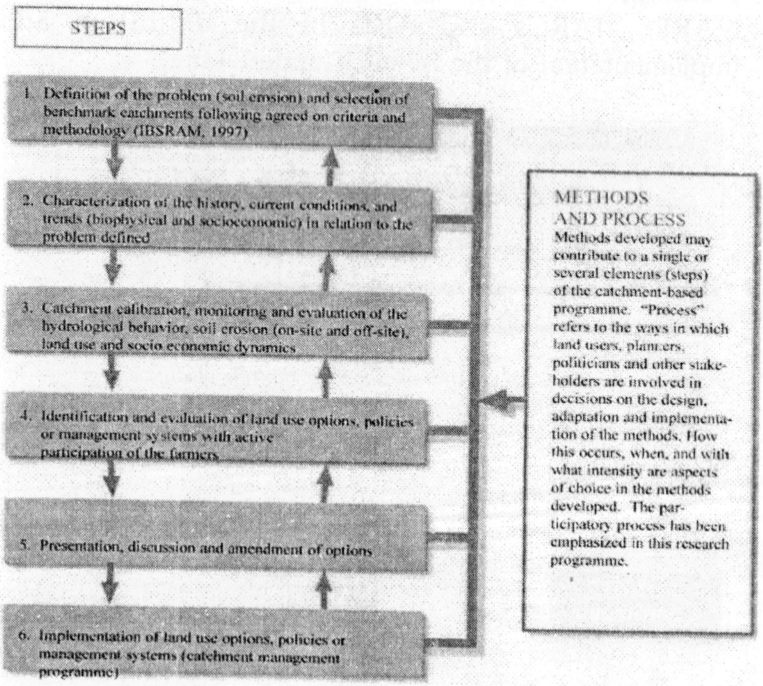

Figure 4. Steps followed in the catchment-based approach to managing soil erosion research

Rainfall, runoff and sediment yields and other climatic data were monitored from the catchments and subcatchments using the installed equipment. Erosion and runoff in each catchment were recorded both manually using staff gauges and automatically using water level recorders. Samples were collected at time intervals of 2 minutes to 1 hour depending on water discharge peaks.

Bed load sediments, i.e., the sediments trapped in the weirs were collected and weighed after each main rainfall event, or once at the end of the rainy season. Runoff and sediment yield data were computed to obtain yearly means. Runoff was derived from water depth after calibration curves were established in the field. Mean annual suspended sediment concentration was combined with water flux data to assess the annual suspended load, using data interpolation between the sampling periods.

Land use was assessed annually by field surveys. Land use types included: forest (Fo), annual crops (C), fallows or pastures (Fa). Crops associated with conservation practices (Cp) were mainly coffee and agroforestry techniques with annual crops. Teak, eucalyptus tree plantations, and orchards were placed in a single category (O). A preliminary evaluation of the off-site impacts of soil erosion was conducted in Thailand by studying the sedimentation of the reservoir downstream of the benchmark catchment.

The "best bet" land management options were identified and introduced through farmer consultations. With the farmers' participation in the selection process, it is expected that they will continue practising the system, which in the long run will provide better income and less resource degradation. The introduced land management options are evaluated for their acceptability and sustainability, and a wider uptake at the community level is promoted to produce greater impact.

The participatory process in soil erosion management research on a catchment scale was employed since the establishment of the consortium and the design of the research programme. A series of consultation meetings and dialogues among various stakeholders, including the NARES, IARCs, ARIs, NGOs, donors and farmer representatives, was undertaken to agree on the design of the research and the various partners that would be involved.

With the aim of promoting activities based on the principles of participation, interdisciplinarity and collaboration, the NARES identified local institutions and project teams composed of researchers of different disciplines. Within the countries, collaboration among relevant partners has evolved. The organisation of these teams from different institutions and disciplines has enhanced the participatory, interdisciplinary and interinstitutional mechanisms that the consortium advocates. Generally, this arrangement is committed through formal agreements signed between and among institutions. It is intended that this arrangement should optimise the use of scarce resources and enhance the synergy of different experts and institutions.

Environmental Factors in Soil Erosion

The relation between environmental factors and runoff amount (R), bed load (BL), suspended load (SL) and sediment concentration (SC) was determined by correlation analysis. The environmental factors included annual precipitation (P), the precipitation ratio (Pr) between minimum monthly precipitation (Pn) and maximum monthly precipitation (Pm), slope gradient (S), catchment area (Surf) and the areal percentage of each land use type.

Runoff and sediment yield were not significantly related to Surf (Table 1). Runoff (R) was highly correlated to orchard land use (O) (r = -0.87) and slope angle (S) (r = 0.56) and rainfall characteristics. Bed load (BL) was significantly correlated to areal percentage of annual crops (C) (r = 0.83). Suspended sediment load was mainly related to C (r = 0.76) and the ratio between minimum and maximum monthly precipitation (Pr) (r = 0.61).

Under these sloping land conditions, the percentage of annual crops not associated with conservation practices appears to be the main factor controlling sediment yield, both in terms of suspended load and bed load, regardless of the sise of the catchment. These results indicate that the area under annual crops seems to be the key parameter related to bed load and suspended load production. Surprisingly, no significant relation was found between runoff coefficient or sediment yield variables and the areal percentages of land occupied by forest or crops associated with conservation practices.

Table 1. Correlation coefficients between environmental factors, run-off and sediment yield variables

	Surf	P	Pm	Pr	S	C	Fa	Cp	O	Fo
R	0.04	-0.95*	-0.91*	0.95*	0.55*	0.25	0.53	0.21	-0.87*	0.39
BL	-0.29	-0.16	0.01	0.22	-0.37	0.83*	-0.4	-0.07	-0.27	0.04
SL	-0.08	-0.54	-0.3	0.61*	-0.28	0.76*	-0.37	0.43	-0.47	0.21
SC	0.33	0.35	0.2	-0.39	0.23	-0.04	0.18	-0.45	-0.05	0.21

Surf (ha) is the catchment area; P, the annual precipitation; Pm, the maximum monthly precipitation; Pr (%) the ratio between Pn, the minimum monthly precipitation and Pm; S (%), the slope; The areal percentages of annual crops (C), fallows or pastures (Fa), crops with conservation practices (Cp), orchards (O) and forests (Fo); R, the runoff ratio; BL, the bed load; SL, the suspended sediment load; SC, the sediment concentration.

Figures 5, 6 and 7 present the relationships between observed and predicted values for mean annual runoff, mean annual suspended sediment load and mean annual

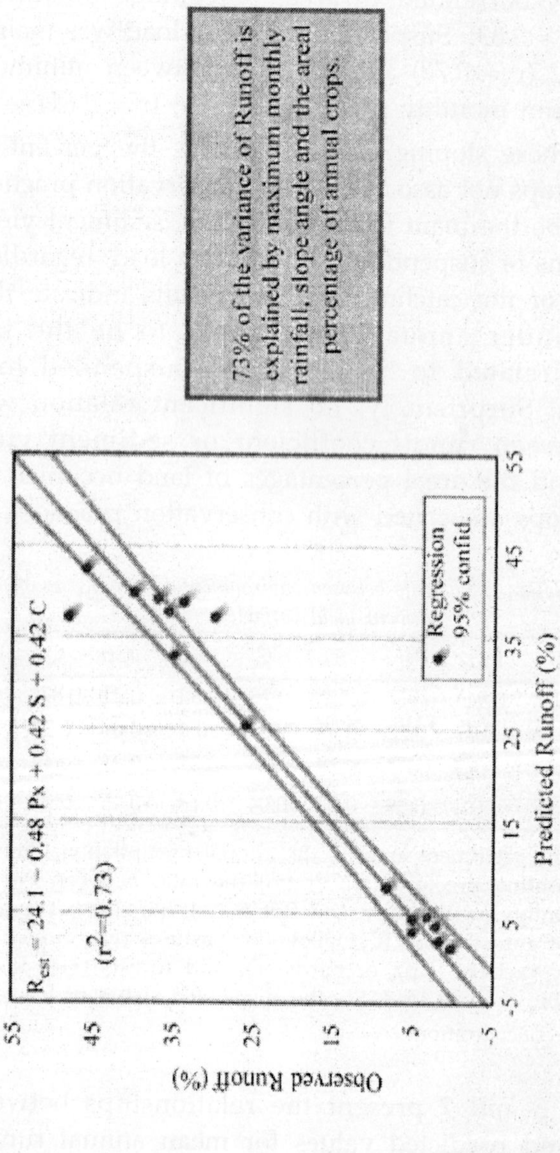

Figure 5. Observed mean annual runoff as a function of predicted runoff or a data set of 16 catchments in Southeast Asia

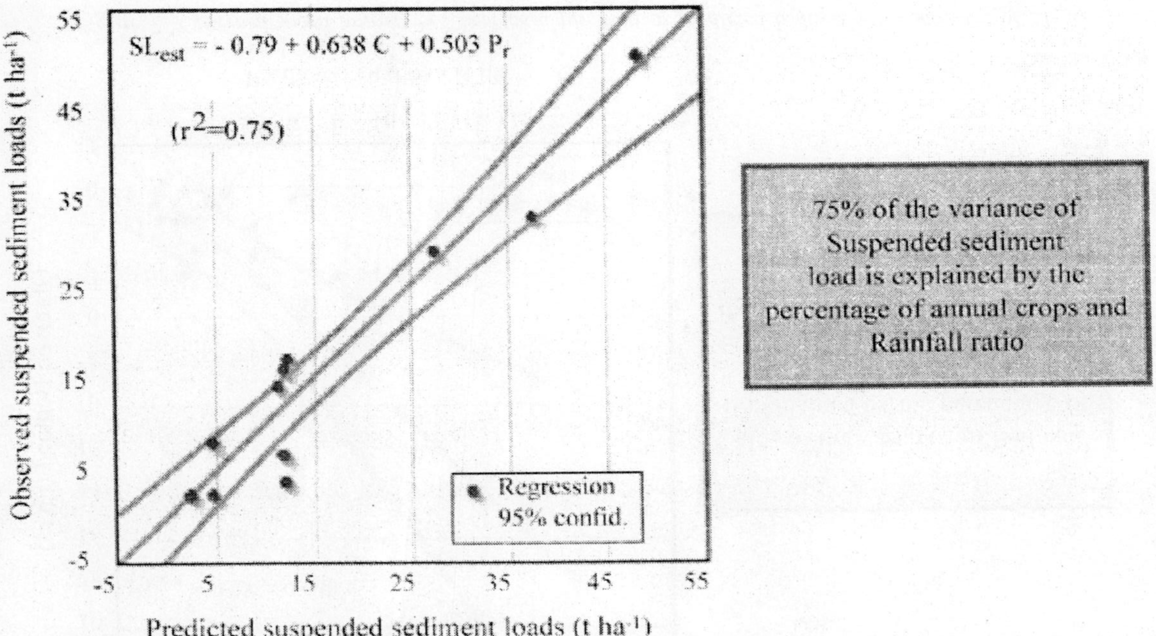

Figure 6. Observed mean annual suspended sediment load as a function of predicted suspended sediment load for a data set of 11 catchments in Southeast Asia

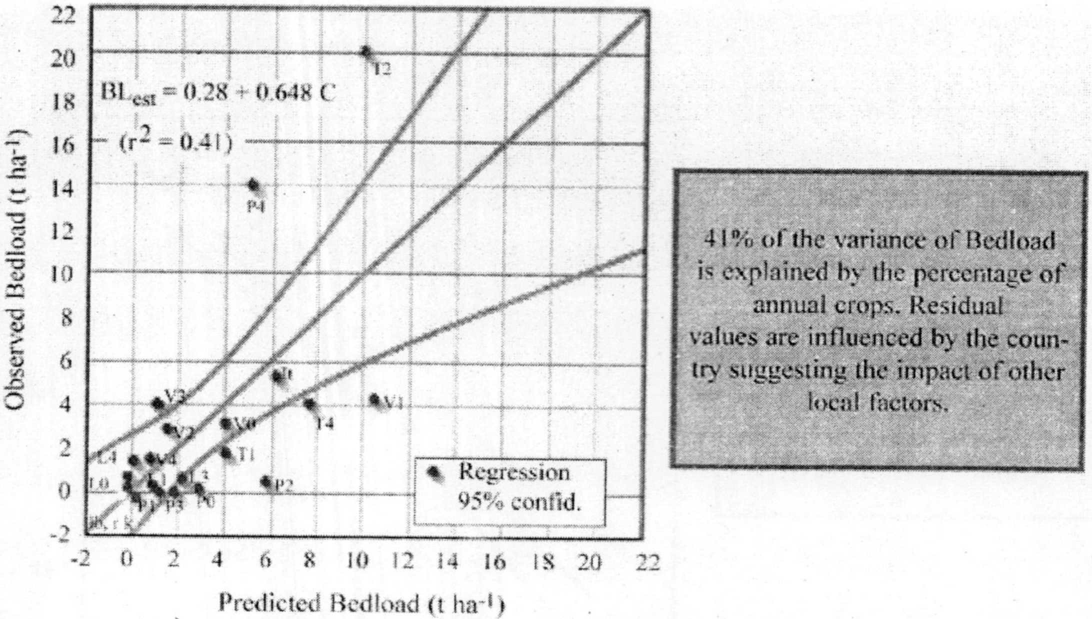

Figure 7. Observed mean annual bed load as a function of predicted bedload for a data set of 21 catchments in Southeast Asia

bed load. More than 70% of the variance was explained by Pr, C and S in the case of runoff, and by Pr and C in the case of suspended sediment load. Less than half of the variance of bed load is explained by C ($r^2 = 0.41$). It is interesting to note that the bed load deviation errors could be ascribed to country conditions, suggesting some site specificity.

Soil Erosion and Land Use

Current land management practices had some effect on the degree of soil erosion in the different subcatchments within the catchments. Except for the results obtained in Nepal, the areas more intensively cultivated to upland crops produced more soil loss than those grown to perennials or left under grass cover (Tables 2 and 3).

Table 2. Observed soil erosion rates in the subcatchments, 2001

Subcatchment	Area	Land Use (ha)	Soil Loss ($t\ ha^{-1}$)
Indonesia			
Tegalan	1.1	50% annual upland crops, coffee and nutmeg on the upper slopes	20.0
Rambutan	0.9	Rambutan and some bare plots	0.8
Kalisidi	20.0	Rambutan	1.0
Laos			
S1	1.7	69% rotating land, 31% teak	0.5
S2	29.3	76% rotating land, 6% upland rice	0.6
S3	19.8	80% rotating land, 12% forest	0.0
S4	27.7	61% rotating land, 11% job's tears, 10% forest, 7% upland rice	2.1
S5	13.1	53% rotating land, 35% forest, 8% upland rice	2.8
S6	2.5	56% rotating land, 13% forest, 31% teak	2.0
Nepal			
W2	72.6	Mixed (45% upland, 5% lowland, 20% shrub, 30% forest)	0.1

Contd. . .

Contd. . .

W3	39.6	Mixed (60% upland, 10% shrub, 30% forest)	0.1
W4	11.5	Mixed (23% upland, 2% lowland, 35% shrub, 40% forest)	0.1
W5	1.6	Upland cultivated (100%)	Traces
Philippines			
MC1	24.9	20% cultivated, 80% Falcata, grassland	0.1
MC2	17.9	40% cultivated, 60% grassland/forest	0.7
MC3	8.0	10% settlement, 15% cultivated, 75% natural grass	1.0
MC4	0.9	40% cultivated, 60% grassland	53.9
Thailand			
W1	11.6	47% soybean-mung bean, 47% tamarind	0.1
W2	9.8	78% soybean-mung bean, 13% shrub	1.6
W3	3.2	94% tamarind, shrub	1.0
W4	7.1	51% soybean-mung bean, 23% mango, tamarind	0.4
Vietnam			
W1	4.8	67% monoculture cassava, 33% natural grass	
W2	9.4	24% cassava intercrop, 59% cassava monoculture, 17% natural grass	
W3	5.2	100% cassava intercrop	
W4	12.4	26% cassava intercrop, 74% natural grass	

Table 3. Observed soil erosion rates in the subcatchments, 2002

Subcatchment	Area	Land Use (ha)	Soil Loss (t ha-[1])
Indonesia			
Tegalan	1.1	50% annual upland crops, coffee and nutmeg on the upper slopes	10.2
Rambutan	0.9	95% rambutan, 5% shrubs	0.2
Kalisidi	13.0	Rambutan	7.8
Babon (Flume)	285.0	All above + rice fields	6.1
Laos			
S1	19.6	49% annual crops, 36% fallow, 14% forest	0.5

Contd. . .

Contd. . .

S2	32.8	67% annual crops, 19% fallow, 12% forest	0.0
S3	51.4	27% annual crops, 48% fallow, 8% forest	0.9
S4	60.2	8% annual crops, 51% fallow, 32% forest	1.9
S5	63.0	2% annual crops, 51% fallow, 14% forest, 34% teak	0.0
S6	0.6	55% annual crops, 6% fallow, 30% forest	0.4
S7	0.6	79% annual crops, 21% forest	4.7
S8	0.6	40% annual crops, 60% forest	1.9
S9	0.7	24% annual crops, 63% fallow, 13% forest	0.1
Nepal			
W2	72.6	Mixed (45% upland, 5% lowland, 20% shrub, 30% forest)	0.1
W3	39.6	Mixed (60% upland, 10% shrub, 30% forest)	0.2
W4	11.5	Mixed (23% upland, 2% lowland, 35% shrub, 40% forest)	0.4
W5	1.6	Upland cultivated (100%)	0.0
Whole (W1)	124.3		0.2
Philippines			
MC1	24.9	20% cultivated, 80% Falcata grassland	0.0
MC2	17.9	40% cultivated, 60% grassland/forest	0.1
MC3	8.0	10% settlement, 15% cultivated, 75% natural grass	0.0
MC4	0.9	40% cultivated, 60% grassland	28.3
Whole	84.5		1.1
Thailand			
W1	11.8	47% soybean-mung bean, 47% tamarind	0.1
W2	9.6	68% soybean-mung bean, 13% shrub	0.9
W3	3.2	59% tamarind, shrub, 41% annual crops	1.5
W4	7.1	51% soybean-mung bean, 23% mango, tamarind	1.2
Whole (Flume)	93.2		2.5
Vietnam			

Contd. . .

W1	3.7	67% monoculture cassava, 33% natural grass	1.3
W2	7.7	24% cassava intercrop, 59% cassava monoculture, 17% natural grass	1.9
W3	10.8	100% cassava intercrop	0.8
W4	7.2	26% cassava intercrop, 74% natural grass	0.6
Whole	45.5		0.5

In Indonesia in 2001, sediment yield was highest in the Tegalan subcatchment dominated by upland annual crops yielding a soil loss of 20 t ha^{-1} in one year of observation. This is presumably because of minimal soil surface litter and little canopy cover of the catchment. On the other hand, the other subcatchments (Rambutan and Kalisidi) planted to perennials (primarily rambutan), lost only about 1 t ha^{-1} of soil during the same period, yielding most sediment during the middle part of the rainy season (January).

In 2002, there was again a relatively higher erosion rate of 10 t ha-1 in the Tegalan subcatchment. However, there was a significant reduction in the erosion rate compared with 2001. This could be attributed to the introduction of fodder grass to reduce erosion and improve farmers' income through livestock integration. It should be noted that in 2002, more than 60% of the area adopted this intervention.

In the Philippines, the smallest subcatchment (MC4), which has a higher percentage of cultivated area, gave the highest soil loss of 54 t ha^{-1}. The lowest soil loss of 0.1 t ha^{-1} was from the MC1 subcatchment which has a lower percentage of cultivated land and a larger area covered with grasses. The MC3 subcatchment, which has the lowest percentage of cultivated area but with 10% of settlements, yielded a higher soil loss. The relatively higher soil loss in this subcatchment may be attributed to erosion from the foot trails and road network. Using a simulation model, Ziegler et al. showed that roads generate runoff sooner during a rainfall event, and have greater discharge values than other surfaces. Sediment transport was also greater. Footpaths emerged as important sources of accelerated runoff from agricultural fields that would otherwise require large amounts of rainfall to produce runoff.

In Vietnam, W1 (predominantly cassava monoculture with some natural grass) produced the largest soil loss of about 5.2 t ha^{-1}. The least was from W4 (predominantly natural grass and cassava intercropping) at 2.0 t ha^{-1}. The larger soil loss from W1 (primarily cassava monoculture) compared with W3 (all cassava intercropping) shows the effect of intercropping on soil losses. At its peak growth, cassava provides only about 47-56% soil cover, whereas mixed cropping or intercropping can increase this protective cover. The effect of natural grass in the subcatchments on soil losses was also evident. Natural grass enhances infiltration, reduces runoff and runoff velocity, and consequently reduces soil loss.

In Laos, the subcatchment with the smallest proportion of rotating land and with some 8% of upland rice (S5) gave the highest soil loss of 2.8 t ha^{-1}. No erosion was observed in the subcatchment with the largest proportion of rotating land and about 12% forest (S3). In Nepal all sites gave very low soil losses in both years, irrespective of land use.

In Thailand, the observed soil losses in 2001 and 2002 were not alarming, being less than 2 t/ha. The increase in soil loss in W3 from 2001 to 2002 was attributed to the change in land use. In 2001, this sub-catchment had more than 90% tamarind and shrub. Most of the tamarind trees were cut in 2002 and soybean was planted, increasing the area planted to annual crops from 3 to 41%. Another cause of variation in erosion can be the rainfall characteristics between years. In 2002, total rainfall was lower and more widely distributed. In 2001, there was a strong rainfall event which accounted for more than 70% of the total runoff.

Soil Erosion and Slope

Rainfall simulation studies in Thailand showed that infiltration rate increased while runoff coefficient decreased with slope gradient. Moreover, sediment concentration and soil detachment decreased sharply with increasing slope indicating that soil erosion decreases with slope for the particular conditions encountered in these studies (Figure 8). These results conflict with the general finding that more runoff is generated from steeper slopes. It is suggested that for convex landforms, the steep midslope zone can act as an infiltration trap for runoff water from the upper gentler zone. This may have substantial impacts on flow volume generated from small watersheds and on water quality. As similar results are obtained under natural conditions in Laos, it can be concluded that they may be valid for some mountain soils with highly stable microaggregates. The data should encourage more research on the physical processes involved in soil erosion on steep slopes.

Tillage Erosion

Tillage erosion results from land preparation and repeated weeding operations. In northern Laos, in a field with a mean slope of 60%, soil losses due to tillage erosion are of the same order of magnitude as those due to water erosion (4-5 t ha^{-1} yr^{-1}). These soil losses affect the most fertile soil layer. Soil clods detached by tillage operations accumulate in depressions or at field boundaries. Because of increasing weed pressure, tillage

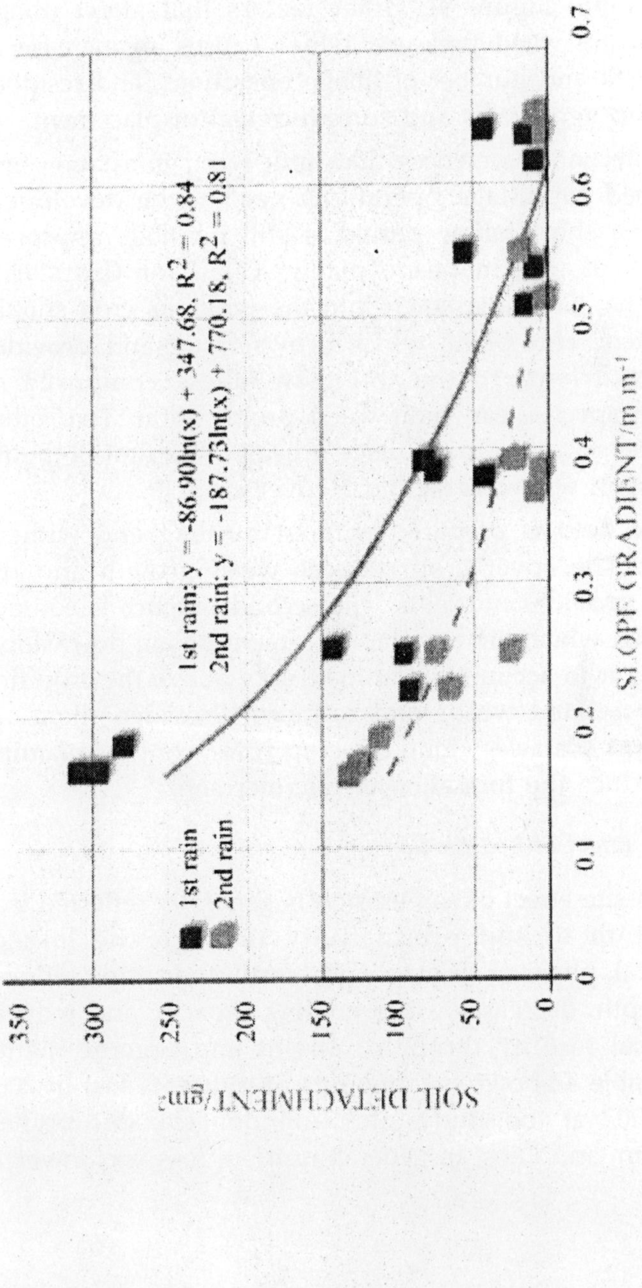

Figure 8. Soil detachment as affected by slope

erosion is likely to become very serious, especially on the steeper slopes, because soil losses from tillage increase with slope (Figure 9). Other factors that affect tillage erosion are weed pressure, which causes an increase in the depth and number of tillage operations, and residues that trap aggregates and so reduce soil displacement.

Alternative practices that limit weed infestation and the need for tillage operations need to be developed. Where a short fallow period is still possible, improved fallows of pigeon pea (Cajanus cajan) or Crotalaria, Leucaena, Gliricidia sown into the previous crop should be tested. This plant, which grows faster and provides more nutrients to the soil than fallow composed of natural species, can limit weed proliferation. Two other practices can also limit tillage erosion. The first is the no till system with cereals and a cover crop.

The field is prepared without burning and without tillage. The cover crop reduces weed growth and the mulch provides nutrients. The second practice is contour planting, which hinders the movement of soil downslope, causing it to accumulate immediately above the crop line and associated weeds and residues. Provided there are sufficient obstacles along the crop rows, contour planting will induce the formation of microterraces.

On-site Effect of Soil Erosion

The on-site effect of soil erosion is generally reflected in a reduction of soil productivity through the loss of chemical, physical and biological fertility. Erosion reduces soil depth, decreases water holding capacity and reduces chemical fertility through nutrient and organic matter loss. Table 4 shows the amounts of nutrients lost in 2001 and 2002 at the study sites in Indonesia, Philippines, Vietnam, and Laos. In general, nutrient loss was lower in

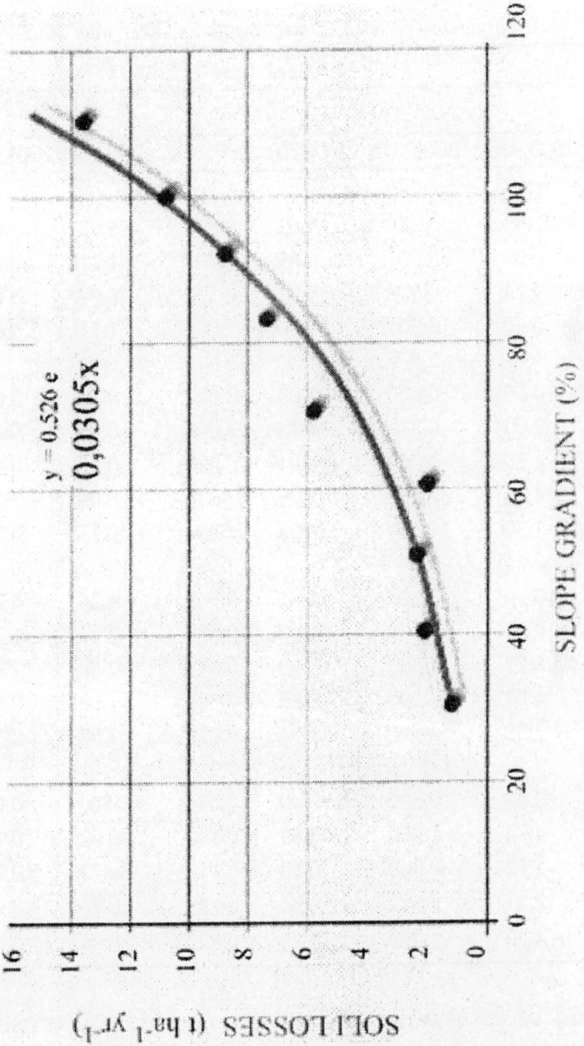

Figure 9. Tillage erosion due to hoeing as influenced by slope

2002 than in 2001, which is consistent with the lower rainfall in that year.

Table 4. Nutrient losses at MSEC catchments in 2001 and 2002

Catchment	Nutrient Loss (kg ha^{-1} yr^{-1})					
	N		P		K	
	2000/01	2001/02	2000/01	2001/02	2000/01	2001/02
INDONESIA						
Tegalan	21.53	5.340	5.82	1.820	9.02	1.650
Rambutan	0.89	0.001	0.89	0.004	1.11	0.002
Kalisidi	9.24	0.028	0.21	0.132	5.97	0.077
Parshal Flume	0.60	0.001	0.00	0.0001	2.11	0.011
PHILIPPINES						
MC 1	0.50	0.000	0.003	0.000	0.50	0.000
MC 2	2.30	0.210	0.0003	0.000	0.05	0.025
MC 3	4.80	0.000	0.004	0.000	0.19	0.000
MC 4	144.20	37.590	0.079	0.0046	6.09	3.319
Whole	1.30	1.59	0.0008	0.0004	0.15	0.166
VIETNAM						
W 1	10.79	3.500	4.81	0.568	4.26	4.730
W 2	10.83	4.600	4.97	0.830	2.46	1.743
W 3	8.73	1.600	3.99	0.393	2.68	1.411
W 4	4.03	1.100	2.25	0.306	1.38	0.830
MW	3.55	0.900	1.94	0.218	2.58	0.747
LAOS						
S 0	0.03	0.000	0.00	0.004	0.00	0.000
S 1	4.74	1.250	0.89	0.196	0.82	0.124
S 2	5.12	0.060	0.93	0.013	0.79	0.000
S 3	12.51	3.314	1.91	0.153	0.76	0.008
S 4	16.27	2.580	2.73	0.122	0.98	0.066

One method of estimating the on-site cost of soil erosion is to calculate the equivalent cost of replacing nutrients lost by erosion. On this basis, the on-site cost of erosion varies among countries depending on soil loss and cost of fertilisers. In 2002, the cost of lost nutrients was highest in the MC4 subcatchment in the Philippines at US$27 per hectare per year (Table 5). This was followed by the subcatchments W1 in Vietnam and Tegalan in Indonesia. The MC4 subcatchment had the highest N loss and the second highest K loss. W1 in Vietnam gave the highest K loss, while Tegalan in Indonesia had the highest P loss.

Table 5. Nutrient losses and on-site costs at MSEC catchments, 2002

Catchment	Nutient Loss (kg $ha^{-1}yr^{-1}$)			On-site Cost (US$ $ha^{-1}yr^{-1}$)
	N	P	K	
INDONESIA				
Tegalan	5.340	1.820	1.650	8.48
Rambutan	0.001	0.004	0.002	0.01
Kalisidi	0.028	0.132	0.077	0.34
Parshal Flume	0.001	0.0001	0.011	0.02
PHILIPPINES				
MC 1	0.00	0.000	0.000	0.00
MC 2	0.21	0.000	0.025	0.16
MC 3	0.00	0.000	0.000	0.00
MC 4	37.59	0.0046	3.319	27.10
Whole	1.59	0.0004	0.166	1.18
VIETNAM				
W 1	3.50	0.568	4.730	9.54
W 2	4.60	0.830	1.743	6.52
W 3	1.60	0.393	1.411	3.55
W 4	1.10	0.306	0.830	2.30
MW	0.90	0.218	0.747	1.93
LAOS				
S 0	0.000	0.004	0.000	0.00
S 1	1.250	0.196	0.124	0.44
S 2	0.060	0.013	0.000	0.02
S 3	3.314	0.153	0.008	0.95
S 4	2.580	0.122	0.066	0.76

Analysis for 2001 showed the same trend with a cost of US$52 per hectare per year, also in the Philippines. In general, it can be seen that the cost of soil erosion is a small fraction of annual income, which varies from $296 to $1 200 per year. Compared with the high cost of labour for erosion control, the costs of erosion are relatively small, and by themselves will probably be insufficient to motivate farmers to adopt conservation measures.

Off-site Impacts of Soil Erosion

An evaluation of the off-site impacts of soil erosion was conducted by identifying potential economic activities downstream that could be affected by erosion occuring in the upper catchments. A simple evaluation of the off-site effect of erosion was carried out at the Philippine site by estimating the cost of dredging the irrigation canals and diversion of the Manupali River Irrigation System. A total of 84 685 m3 of sediments has been estimated to have been transported into the system since 1995. Assuming that 0.5% comes from the Mapawa site, it was estimated to have contributed 423 m3 of sediments to the irrigation system or an equivalent of US$250 in terms of the cost of dredging.

In Thailand, the sedimentation of the reservoir downstream was also evaluated. The amount of sediment that had accumulated in the reservoir since the construction of the dam was determined by comparing the topographic map of the site before the operation of the reservoir, and the bathymetric map prepared by conducting a survey before the rainy season of June 2002.

Siltation in the Mae Thang reservoir at the study site in Thailand showed a 10% reduction in storage volume after seven years of operation. The soil erosion rate calculated from the volume of sediments that has accumulated in the reservoir was 51 t ha^{-1} yr^{-1}. At this rate of erosion and sedimentation, the life of the reservoir would not be more than 70 years. The rate may be an overestimation, but the value is close to that presented by Inthasothi et al.

In addition to the effect on the life of the reservoir and its irrigation service area, erosion in the upper catchments may also cause an accumulation of heavy metals and other toxic substances in the reservoir. Granulometric and chemical (heavy metals) analysis made on lake sediment samples showed no significant pollution. However, fish farming may be affected particularly during the first heavy rains due to the accumulation of pesticides used in the watershed.

Options of Land Management

The recommended land management options were identified in consultation with the farmers and introduced in 2001 and 2002. In most instances, the options introduced in the catchments were variants of contour hedgerow farming combined with improved soil fertility management and animal production.

In the Philippines, natural vegetative strips (NVS) being promoted by ICRAF were introduced in 2002 with emphasis on soil fertility management as suggested by the farmers. This has been done using native grasses and some agroforestry crops as hedgerows. It is worth noting that several farmers have already adopted this technique as a result of promotion activities by ICRAF in the area.

Adoption seems to be affected by the farmers' tenure system. About half of the landowners, but no tenant farmers, have adopted some conservation measures so far. The major reason for the lack of adoption by farmers interested in the practice is the cost of establishment.

In Indonesia, the conversion of arable land into fodder banks of Elephant grass (Pennisetum purpureum) combined with livestock production was first implemented in November 2001, and completed in 2002 by the procurement of cattle. The grass was planted on alternate terraces of land currently used for annual upland crops. In terms of the severity of erosion, the area planted to annual upland crops needs priority attention.

Table 6. Impact of innovative farming technologies on bedload, suspended sediment load and total sediment yields, Houay Pano catchment, Laos. Traditional system includes fallow land with slash and burned fields

Ttraditional and innovative farming systems	Catchment area	Bedload	Suspended Load	Total Sediment Yields
	ha	t ha^{-1}	t ha^{-1}	t ha^{-1}
Slash & burn (control)	0.62	4.74	0.99	5.74
Improved Fallow	0.64	0.4	0.01	0.42
Improve Fallow + Contour planting	0.567	1.95	0.56	2.51
Mulch + No tillage	0.727	0.11	0.47	0.58

Table 7. Normalised labour required (days ha^{-1}) for the cultivation of upland rice and Job's tears, Houay Pano, Laos, 2002

Operation	Slash and Burn	Improved fallow	Contour Planting & Improved fallow	No tillage
Field preparation	41	43	71	29
Burning	1	1	1	
Second clearing	26	57	29	63
First weeding	36	28	18	21
Second weeding	38	2	6	
Tillage Sub-toal	142	131	125	113
Herbicide				5
Transport/Planting pineapple			26	
Planting main				

Contd. . .

Contd. . .

cereal crop	43	34	33	21
Planting cover crop		42	30	6
Harvest and transport	25	6	24	4
Grand Total	210	171	182	138

The fodder grass is expected to reduce erosion and serve as feed for the livestock. The identification of this option was based on the lesson learned that farmers' adoption and improvement of conservation measures is determined by the economic contribution of the measure to the household economy. Farmers are attracted to a practice only if the practice promises an economic benefit, and so this consideration must be put forward in the participatory technology selection process.

In Vietnam, Tephrosia candida and vetiver grass (Chrysopogon zizanioides) have been used since 2001 as hedgerows combined with the use of a high-yielding variety of cassava. Possibly as a result of a nearby demonstration site on alley cropping, the farmers believe that the system will reduce runoff and soil loss, add organic matter and improve soil fertility through the addition of hedgerow trimmings.

In Lao PDR, the traditional practice of slash and burn has been closely monitored over a period of two years, and "improved" options are currently under observation. These include (i) improved fallow involving later intercropping of pigeon pea and Crotalaria, (ii) contour planting of Job's tears (Coix lachryma-jobi) with pigeon pea, or Crotalaria and pineapple (Ananus comosus), and (iii) no tillage with the use of herbicides.

In terms of soil erosion, these options yielded 0.4, 2.5 and 0.6 t ha^{-1} of soil loss respectively, compared with 5.7 t ha^{-1} from the traditional slash and burn system (Table 6).

The cultivation of upland rice and Job's tears also required more labour of 210 days per hectare in the slash and burn system compared with 171, 182 and 138 days per hectare for improved fallow, contour planting, and no tillage, respectively (Table 7).

Bibliography

Batey, T., "Soil husbandry", *A practical guide to the use and management of soils,* Soil and Land Use Consultants, Aberdeen, 1988.

_______________., "Soil structure and the production of arable crops", *Journal of the Royal Agricultural Society of England,* 1971.

Bos, J.F.F.P. and J. de Wit, "Environmental impact assessment of landless monogastric livestock production systems", *International Agricultural Centre,* Wageningen, The Netherlands, 1996.

Davies, D B; Payne, D, "Management of soil physical properties", *Russell's soil conditions and plant growth. Wild,* A. *ed.* Longman, Harlow, 1988.

De Haan, C., Steinfeld, H. and Blackburn, H., "Livestock and the Environment: Finding a balance", *Draft Summary Report,* FAO/World Bank, 1996.

Duxbury, J.M., Harper, L.A., and Moiser, A.R., "Contribution of agroecosystems to global climate change. In: Agricultural Ecosystem Effects on Trace Gasses and Global Climate Change", American Society of Agronomy Special Publication Number 55, Madison, WI, USA, 1993.

FAO, "A framework for land evaluation", *Soils Bulletin* 32, FAO, Rome, 1976.

_______________, "Forest resources assessment 1990; global synthesis", *Forestry Paper,* FAO, Rome, 1995.

Greenland, D.J., "Soil management and soil degradation", *Journal of soil science,* 1981.

Greenwood, P B., "The effects of subsoiling on soil physical properties and crop production", *PhD Thesis*, Lincoln College, University of Canterbury, 1989.

Raghavan, G S V; Alvo, P; McEyes, E., "Soil composition in agriculture: a view toward managing the problem", *Advances in soil science,* 1990.

Shepherd, T.G., "Sustainable soil-crop management and its economic implications for grain growers", *The Proceedings of the International Conference on Sustainable Land Management,* Napier, Hawke's Bay. Henriques, P R *ed.* Hawke's Bay Regional Council, Napier, 1992.

_______________ ., "Towards sustainable soil and crop management of arable land", *Technical Record No. 56.* DSIR Land Resources, 1991.

Sims, D., "META: A New Approach", *AGL Land and Water Newsletter,* No. 26, August, 1986.

Index